Dynamic Stability Control of Narrow Tilting Three-Wheel Vehicles

By
Sukumar

Chapter 1

Introduction

Narrow tilting three wheeled (NTTW) vehicle is a promising personal mobility platform which is a practical alternative to four and two wheeled vehicles. Figure 1.1 shows a schematic of NTTW vehicle with two front wheels while taking a turn, where the body, as well as all the wheels lean to a side. It provides comfort and safety as a four wheeler and has a low carbon footprint like a two-wheeler, thus making it a combination of positive characteristics of a four and two-wheeler [1]. The motivation of this research is directed towards the current personal mobility challenges. During the last ten years, a lot of research has gone into the development of narrow tilting three wheeled (NTTW) vehicle technology as a viable alternative to four and two wheelers. However, there are many challenges involved in order to make this vehicle safe, comfortable and acceptable to the consumers. Due to narrower track width and high center of gravity, the dynamically unstable NTTW vehicle is made to lean towards the direction of turning in order to maintain stability while maneuvering. Hence, along with the conventional vehicle dynamics which consist of three translational motions (i.e. lateral, longitudinal and vertical), there is a need to consider an additional degree-of-freedom due to leaning of the vehicle body to achieve stability of NTTW vehicle. Tilt dynamics for a narrow vehicle is concerned with degree of steering while tilting during cornering maneuvers. Hence, the design of an integrated vehicle stability control system under various driving conditions such as rapid maneuvering, slippery road surfaces for various speeds and accelerations becomes a challenge. As a result, significant research has been directed towards design and development of tilt control strategies for such vehicles. Further, the associated research is directed towards achieving low carbon or green vehicle configuration in the form of electric/hybrid-electric power-train.

The goal of this research work is to address vehicle dynamics stability issues typical to the proposed platform. It is aimed at analysis of vehicle dynamics along longitudinal (braking, traction control), lateral (steering maneuvers, yaw control) and vertical directions individually. Furthermore, the inter-relationship among the aforementioned functions is analyzed and an optimized integrated control strategy for active control of vehicle dynamics under various operating conditions is proposed. The implementation of such integration poses some interesting engineering challenges to maintain currently accepted levels of ride, handling and stability performances.

(a) Side View

(b) Front view while tilting

Figure 1.1: Schematic diagram of an NTTW vehicle

1.1 Research Motivation and Objectives

The NTTW vehicle is a viable alternative for a four wheeler and has immense potential, but there are many challenges involved in order to design this vehicle as safe, comfortable and acceptable to the customers. As compared to a standard vehicle, the track width of a narrow vehicle is much less resulting in an increase in the ratio of center of gravity to track width. This poses stability problem for a narrow vehicle as there is a need to lean the vehicle body towards inside while cornering. Tilt dynamics for a narrow vehicle is concerned with how much to steer and simultaneously how much to tilt while cornering. Also, the challenge is design and development of an automatic tilting balance control system under driving conditions like rapid maneuvers, slippery road surfaces, slanting road surfaces for all speeds and accelerations/decelerations. Thus, observations regarding limitations in existing knowledge have defined a strong motive to pursue work in this direction. The objectives of the subject research work were set as follows.

i. To formulate a mathematical model of narrow tilting three wheeled vehicle

ii. To implement the model and analyze vehicle dynamics using tools such as Simulink

iii. To investigate how longitudinal, lateral, roll dynamics are affected by tilt mechanism and to assess stability functions that are feasible to integrate for the overall control design implementation

iv. To evaluate motor requirements for the safe operation of the vehicle within the operating region

v. To formulate an integrated control strategy to achieve optimum vehicle dynamic behavior under various operating conditions

vi. To verify control logic for standard maneuvers using hardware-in-the-loop set up

1.2 Gap analysis

Majority of the research, as observed from the literature survey, is centered on modeling of NTTW vehicles to investigate various control strategies for roll-over prevention and tilting during cornering. The challenge involved in the coupling of tilt dynamics with other aspects of vehicle motions such as acceleration/deceleration while cornering are not fully addressed.
It is noted from relevant literature that there is a scope of research to design and implement integrated chassis dynamics control (lateral-tilt and longitudinal dynamics) for the NTTW vehicle. Additionally, there is a scope to enhance the stability of NTTW vehicle by utilizing electric in-wheel motor as powertrain. Such an integrated control system to provide dynamic stability for electric NTTW vehicle could not be found in the literature. There is a need to further investigate the application of modern control techniques to enhance the dynamic stability performance of NTTW vehicle.

1.3 Research Methodology

The thesis investigates the integrated dynamics of the NTTW vehicle by the formulation of seven degree of freedom (DoF) nonlinear mathematical model. Then different dynamics stability control strategies are analyzed in a systematic manner and an optimized integrated chassis dynamics control system for NTTW vehicle is proposed. The approach was to study multivariable controllers for integrated vehicle dynamics to achieve the desired performance of the proposed NTTW vehicle. The investigations were carried out by examining different control strategies consisting of basic classical control scheme (based on PID controller), optimal control scheme and sliding mode control scheme. These control strategies were analyzed step by step. In addition, an observer based on EKF technique was designed to reconstruct the unmeasured state of lateral-tilt dynamics from the measured states. Finally, the most suitable robust controller is ascertained and thereafter validated through hardware-in-the-loop simulation. Subsequently, the findings of the research work are reported.

1.4 Thesis Outline

The documentation of the thesis work has been outlined in this section.

Chapter 1: Introduction

This chapter gives introduction to the subject of the research work. The motivation of this research is directed towards the current personal mobility challenges in the form of traffic congestion, depleting parking room and high emission. The upcoming technology of narrow tilting three wheeled (NTTW) vehicle for two passengers has been the subject of research work in demand as the work would be useful to the society. It also presents the research objectives and methodology of the conduct of this work.

Chapter 2: Literature Review

This chapter reports the review of pertinent literature available on NTTW vehicle and its control dynamics. First, few important NTTW vehicles developed worldwide are described with salient features. Then, the vehicle platform dynamic control systems are explained in brief with an emphasis on recent developments in integrated vehicle dynamics control systems. Thereafter, NTTW vehicle tilt dynamics, mechanism and control systems are reported. With the advent of electric drives, the options for prime movers are increased. Comparison of electric/hybrid vehicle were made. Lastly, NTTW vehicle handling characteristics are explained.

Chapter 3: Mathematical Modeling of NTTW Vehicle and its Parametrization

This chapter explains the steps undertaken during formulation of mathematical modeling of NTTW vehicle. A seven DoF non-linear detailed mathematical model of an NTTW vehicle chassis dynamics along longitudinal, lateral and tilt (of body) motions is formulated using first principle method. It includes the consideration of typical forces acting at tyre-road interaction point and also other dynamic forces encountered during various maneuvers of such a vehicle. A realistic design of specifications for NTTW vehicle has been chalked out and the design parameters are also given (detailed explanation given in Appendix A).

Chapter 4: NTTW Vehicle Integrated dynamics Control Design: Problem Formulation

The control design problem has been defined in this chapter. It presents the definition of certain crucial parameters required in analyzing the handling behavior of NTTW platform, such as static envelope and perceived acceleration. An overall concept of the integrated vehicle control strategy is provided wherein it is posited that two different controllers for lateral-tilt and longitudinal dynamics are needed to obtain an integrated stability controller. A set of control objectives are drawn based on the functionality of the NTTW platform and typical vehicular stability conditions. Then, a set of test maneuvers are reported which are considered to analyze the proposed controllers in chapters 5 through 10.

Chapter 5: Design of Baseline Controller Strategy

As a first step towards the design of a dynamic stability controller for the NTTW vehicle, the fundamental issue of tilt stability while maintaining the desired driver maneuver are discussed in this chapter. An integrated controller is designed using classical PID control strategy as they are extensively employed in the industry. This chapter describes details of the baseline controller strategy and its design. Initial simulation studies of the baseline controller are presented and are used in later chapters for reference. The overall response of the baseline controller shows that the desired trajectory is not followed satisfactorily and use of advanced control strategies are essential for such a system.

Chapter 6: Design of Multivariable Tilt and Drive (MVTD) Controller Strategy

In this chapter, a multivariable controller strategy using a state feedback approach (linear quadratic regulator) is implemented. This optimal control technique is designed for the integrated dynamics stability control of the NTTW vehicle. The term 'integrated' signifies a simultaneous control approach for the lateral-tilt and longitudinal dynamics. The theory of the LQR (linear quadratic regulator) control strategy is first explained in brief and then the design of the MVTD control strategy is explained in detail. A comparison is made on the performances of baseline and MVTD controller strategies by simulating the vehicle model in various operating conditions. These studies and simulations show improved vehicle handling and stability performances with proposed novel strategy.

Chapter 7: Estimation of Unmeasured States for Nonlinear NTTW Vehicle Model

In a practical scenario, all the states are not available for feedback and obtaining full information of the states are limited by their practical utility. Thus, the design of an observer becomes an attractive approach to estimate the unmeasured state of NTTW vehicle. In this chapter, an observer based on extended Kalman filter (EKF) is proposed to estimate the unmeasured state signals using the non-linear system model and the measured state signals.

Chapter 8: Design of Integrated Multivariable Sliding Mode (IM-SM) Controller strategy

This chapter presents a design methodology of a robust integrated multivariable sliding mode control methodology to achieve stability and acceptable handling of the NTTW vehicle. Similar to MVTD controller design, two multivariable controllers based on sliding mode control (SMC) approach are designed, one for lateral-tilt dynamics and the other for longitudinal dynamics. In addition, the states of the lateral-tilt dynamics are modified to ascertain zero steady-state error tracking and thereby a modified SMC controller design is proposed. Further, the stability of the proposed IM-SM controller is presented using Lyapunov analysis. Simulation results are presented to analyze the controller's performance in various test scenarios.

Chapter 9: Comparison of MVTD and IM-SM Controllers

A comparison between the two proposed multivariable controllers, i.e. the MVTD controller and the IM-SM controller is presented. Different test scenarios have been simulated and the corresponding responses are monitored. A detailed investigation is showcased by means of quantitative, such as input power requirements, as well as qualitative analysis, such as tracking response. The responses of each controller are compared in presence of external disturbances such as variation in vehicle resistance forces, additive disturbances in input channel and variations in parametric values.

Chapter 10: Hardware-In-the-Loop (HIL)Testing

In this chapter, a suitable HIL set up is developed to test and validate the proposed stability control systems for the narrow three wheel tilting vehicle. An HIL setup is developed based on a host computer with MATLAB/Simulink tool and a low power embedded ARM M4 MCU based platform from STM microelectronics as controller target board. All three controllers have been programmed on STM board which is a low power embedded CPU based platform. This indicates the feasibility of these complex multivariable controllers for practical deployments on automotive grade embedded ECUs.

Chapter 11: Conclusion and Future Scope of Work

This chapter presents a summary of this thesis work and scope for further work is reported.

Appendix A: Design of Proposed NTTW Vehicle

Based on the study made on various NTTW vehicle developed by automotive manufacturers and academicians, the author has chalked out design specifications. The author has attempted to propose a practical and implementable NTTW vehicle. The parameters used in this work are the combination of mostly realistic values and some educated assumptions. The calculations of the parameters are given in detail in Appendix A.

Chapter 2

Literature Review

2.1 Introduction

After the second World War, small non-tilting three wheeled vehicles gained popularity as economic delivery vans and commuters due to lower weight and size. But due to the problem of handling instability, the average automobile was configured with four wheels. Generally, three wheeled vehicles are designed as cars with one front or one rear wheel [2]. They are compact and with the advent of electric drives, they have become obvious choice as commuters with today's congested road traffic. Hence, they are often compared with their four-wheel counterparts. Dynamic behaviors of the two types of vehicle with and without the use of control system are compared with [3] and concluded that a three-wheeler could be as stable as a four-wheeler. Thus, researchers started giving more efforts to design three wheeled vehicles with more and more stable platforms. Consequently, this resulted in crystallizing the requirements of a small, narrow and tall vehicle that could seat two people. Also, the associated technological challenges were understood in the form of leaning the vehicle body while cornering to provide stability. Hence, the future narrow tilting three wheeled (NTTW) vehicle will differ from their counterparts in their ability to control vehicle roll and therefore improve stability. The authors [4] compared the performance of tilting and non-tilting vehicle having same specifications and observed that the active tilt system improved the maneuverability (turning circle radius) of the vehicle for all longitudinal speeds, especially at higher speeds.

The NTTW vehicle is an upcoming technology. Hence, discussion in this thesis is based on limited past work and few prototypes manufactured for proving concept and usability. Two configurational concepts have been tried out by different researchers. In one configuration, two wheels are placed at the front and one wheel is placed at the rear. In the other configuration, two wheels are kept at the rear and one wheel is kept at the front. Single rear wheel configuration simplifies the design of rear wheel chassis and drive-train, whereas single front wheel configuration is more stable in dynamic performance [5].

As compared to a standard vehicle, the track width of a narrow vehicle is much less resulting in an increase in the ratio of center of gravity to track width. This poses stability problem for the narrow vehicle while cornering. Hence, there is a need to lean the vehicle body towards inside while cornering.

This chapter reports the review of pertinent literature available on NTTW vehicle and its control dynamics. First, few important NTTW vehicles developed worldwide are described with salient features. With the advent of electric drives, the options for prime movers are increased, hence a comparison of electric/hybrid vehicle are made. Then, the vehicle platform dynamic control systems are explained in brief with an emphasis on recent developments in integrated vehicle dynamics control systems. Thereafter, NTTW vehicle tilt dynamics, mechanism and control systems are reported.

2.2 NTTW Vehicle Worldwide

Some of the research work has been published in the literature for NTTW vehicle at Universities by academicians, where prototypes are built and studies are carried out. Work carried out at industries have a limited amount of literature. A large majority of work carried out at Institutions has been centered around the modeling of NTTW vehicle in order to investigate the strategies employed to control tilting mechanisms. Some of the important developments on NTTW vehicle worldwide are described below.

a) **Piaggio MP3 LT 400:** It was marketed in 2009 as a tilting three wheeled scooter, Piaggio MP3 has two front wheels and one rear wheel (Figure 2.1). In 2009, the MP3 LT 400 was launched in the UK with wider front track width (465 mm) as a tricycle (rather than a motorcycle). Also in 2009, a plug-in hybrid version of MP3 was launched in the USA with 125 cc engine and a battery-driven 2.6 kW electric motor [6].

Figure 2.1: Piaggio MP3 Scooter [7]

b) **Carver One:** This vehicle was launched in the European market (named Carver One) in 2007. The Carver vehicle developed by Brink Dynamics [8] has one steerable front wheel and two driven rear wheels. This is designed with 1.3 m of track width, 1.4 m of height and 3.4 m of length. The tilt control is actuated through the hydraulic system. The vehicle has a tilting body module and a non-tilting rear module (Figure 2.2). The driver's input torque is translated into actual steering torque on the front wheel and a tilting torque of the tilt actuator. At low speed, the power steering functions and steering wheel angle follow the driver's intent through the steering wheel. But at higher speed, the power steering becomes non-functional and the driver's intent through the steering wheel is partially converted into steering wheel angle and partially into body tilt angle. They automatically assume the ideal ratio which is suitable for a speed of travel [9].

Figure 2.2: Carver One Vehicle [10]

c) **CLEVER:** The CLEVER (Compact Low Emissions Vehicle for urban transport) is a NTTW vehicle that was developed in a collaboration between the University of Bath, UK and BMW in Europe for city travel by combining the comfort and safety. The maximum weight of CLEVER is 412 kg, the track width is 0.84m and the wheelbase is 2.4m. CLEVER has front wheel, cabin and rear module separately coupled. It has a rear module on two wheels housing engine assembly which does not lean during cornering and the body is made to lean (Figure 2.3). The CLEVER is designed with neutral steering. The rear wheel steering is introduced and directly proportional to the tilt angle of the vehicle body [6]. It has an active tilt control system to stabilize the vehicle during cornering [8]. The tilt control system has two single acting hydraulic actuators, which controls the lean angle of the tilting cabin by rotating it with respect to the rear module. A proportional directional control valve with a closed center position modulates the flow to the actuators.

Figure 2.3: CLEVER Vehicle [11]

d) **i-Road:** TOYOTA brought into Japan and France markets a compact NTTW vehicle for personal mobility named 'i- Road' which is an all-electric, three wheeled rear steered [12]. It seats two occupants in tandem. It is having a length of 2.4 m, height of 1.4 m and width of 0.85 m, with the curb weight of 300 kg (as shown in Figure 2.4). It is powered by two 2 KW hub motors. The left and right front wheels move up and down independently synchronized in response to the driver's steering and automatically select the optimum lean angle when cornering. It has rear steering system which is used during low-speed maneuver and while parking. A CPU manages the degree of lean of each front wheel using inputs from steering angle, gyroscope and vehicle speed sensors. The tilting of the vehicle body is affected by the use of lean actuator fixed to each front wheel's suspension arm. However, technical information available in the literature is limited.

Figure 2.4: Toyota I-road Vehicle [13]

e)Prototype of University of Minnesota, USA: The prototype has two non-driven front wheels and one driven rear wheel. The unladen weight of the vehicle is 225 kg. The overall height is 1.84 m with a wheelbase of 1.39 m. The track width is 1 m. The NTTW vehicle was built by adapting components of a Yamaha scooter chassis and 125 cc engine is used (Figure 2.5). The tilting system is provided with four bar mechanism and the vehicle can tilt up to $\pm 35°$ with a $\pm 10°$ additional bump capability [14].

Figure 2.5: University of Minnesota [14]

f) Prototype of University of Nagoya, Japan: A switching dynamic prototype model of NTTW vehicle was developed and dynamic behavior of the vehicle was studied at the University of Nagoya, Japan [15]. The NTTW vehicle has one front wheel and two rear wheels (Figure 2.6). The two rear wheels are actively tilted by a motor fixed on the rear part. This model considers states like running the vehicle with one rear wheel off the ground beside normal running. The driving stability of NTTW vehicle was analyzed through simulation as well as the actual prototype. Further work on performance analysis and controller design is not yet reported.

Figure 2.6: University of Nagoya [16]

g)Prototype at Torino Polytechnic, Italy: The development of the NTTW was taken up in 2005 at Torino Polytechnic, Italy. It is designed with a gross weight of 300 kg (driver added), width of 1.2 in and leaning capacity up to maximum 45. It has two front and one rear wheel. It is a hybrid vehicle, with front two wheels are driven by electric motors and the rear wheel is driven by IC engine [1] and [17]. The details of experiments carried out on the prototype and model validation processes are presented in [17]. It uses DTC at speed type up to 25 kmph and STC above this threshold speed. A torsional resilient joint is designed to couple/decouple the tilting and steering motion, which is placed between handlebar and front wheels [18]. The variable torsional stiffness decreases as the speed of vehicle increases. The resilient joint is configured to have two stop positions so that steering wheel rotates relative to steerable road wheels within a predefined interval. Once the maximum relative angular position is reached in either rotational directions, steering wheel and steerable road wheels are connected rigidly so that they rotate in equal amount for further rotation. The design of control system allows for non-zero lateral acceleration to provide the feeling of the presence of centrifugal force to the driver.

h)Honda Gyro Canopy Prototype: The CLEVER was marketed in Europe after a good amount of studies on NTTW vehicle were carried out. Many researchers [19], [6] in the University of Bath, UK contributed on experimentation and analysis of the dynamic performance of CLEVER vehicle before the same was put on the market. It may be noted that although the dynamic tilt control using DTC scheme of control was satisfactory, the major problem was the case of rolling of the rear non-tilting part out of the bend during rapid steer input and high levels of lateral acceleration [2]. This was due to large tilting torque demand during the said cases as the reaction forces were applied through the rear non-tilting part. This led to investigate further on the challenges and a new prototype was made and named "Honda Gyro Canopy" in the University of Bath, UK. It was a 69 cc moped which was re-engineered with a tilting front assembly and a rear non-tilting part. Thus, the configuration was similar to that of CLEVER. In this case, STC system was used to control leaning of the vehicle whereas DTC system was used in the CLEVER vehicle. [2]carried out experimentations with STC system and concluded that STC system suites better than DTC as tilting control system for such configuration of NTTW vehicle.

2.3 Power Source for NTTW Vehicle

2.3.1 IC engine powered vehicle

Most of the earlier prototypes and products of NTTW vehicle released into the markets are powered by IC engines. For example, Carver, CLEVER vehicles.

2.3.2 Hybrid electric powered vehicle

The hybrid drive concept appears in more than one forms depending on the combination of the energy sources and propulsion systems used in the vehicle [20]. A hybrid electric vehicle (HEV) comprises of both IC engine and an electric motor. An HEV can benefit from the best features of both conventional internal combustion engine (ICE) and electric drive. There could be three different kinds of HEV, e.g series, parallel and series-parallel. Series HEV has a comparatively simple structure and less complicated energy management strategy. A series HEV provides tractive power to the drive wheels using electric motors and ICE is used to recharge the batteries, which is the power source for the motors. The ICE generally runs in its efficiency region all the time as there is no direct mechanical connection between ICE and vehicle load. The main control strategy in a series HEV is to keep the level of the charge of the batteries above a specified range. Electric motors are the main component of series HEV and AC or DC motors can be utilized. Figure 2.7 shows a layout diagram of a series HEV.

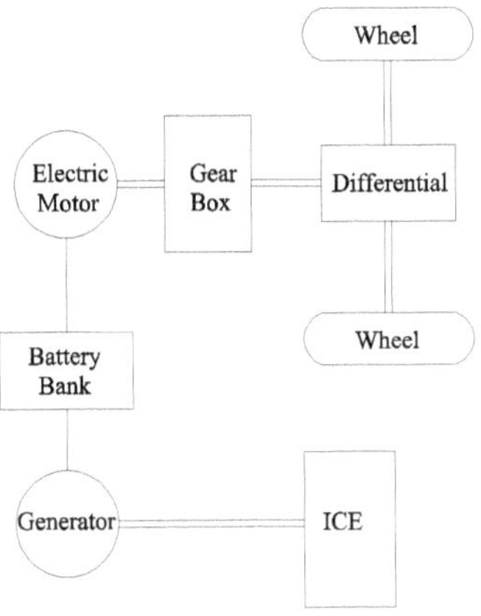

Figure 2.7: Series HEV configuration of a vehicle [21]

The second kind is the most common type of HEV, i.e. the parallel HEV which uses the combination of both IC engine and electric motor [22]. One prototype of NTTW parallel HEV was developed at Torino Polytechnique, Italy [17] and is shown in Figure 2.8. A similar configuration of an NTTW vehicle is proposed in [18]. Parallel HEV operational strategy and system synthesis through simulations are presented in [23]. The parallel HEV, considered in this case, starts in motor only operational mode for optional fuel economy. As the vehicle speeds up, when the operation of the engine becomes beneficial, the motor is not operated. The motor also assists the engine, when the engine cannot meet the driver demand. It also helps in optional engine sizing. The parallel HEV is also provided with controllers for sub-systems in order to have smooth functioning of the vehicle. The electronic control unit (ECU) controls and monitors the dynamic stability of the vehicle. For HEV, the IC engine is employed over a speed of 40 kmph and the electric motor is operated at various operating conditions and also during transient torque deficits created by IC engine. Hence torque controllers need to track the torque of electric motor precisely for desired performance of HEV [22].

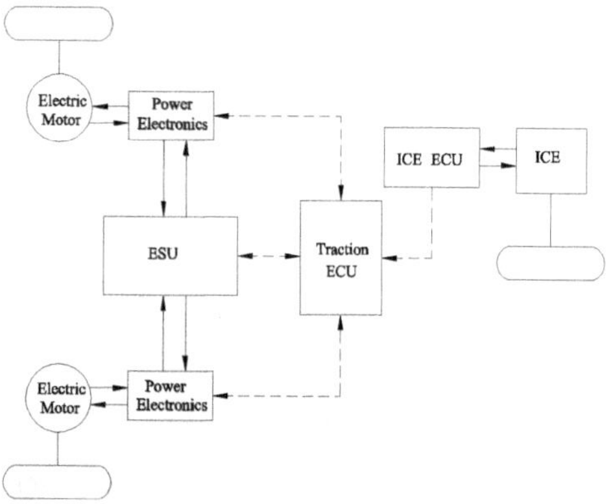

Figure 2.8: Parallel HEV configuration of NTTW vehicle [17]

The third kind is that at low speed, the vehicle is driven by only the electric motor and when the velocity increases, the vehicle is driven by ICE and electric motor. This kind is called the series-parallel HEV.

2.3.3 Electric powered vehicle

With the improvement of motors and batteries, electric vehicles (EV) have achieved enough performance in comparison to IC engine drive vehicles. An EV has distinct advantages. Table 2.1 shows the basic demands of EV drive and their effects on vehicle performances [24].

Table 2.1: Basic demands of EV drives and their effect on vehicle performance

Demands	Effects
High torque at low speeds and Constant power at high speeds	Driving performance
Continuous and smooth drive and control	Comfort
High specific power and specific torque	Vehicle performance
Ruggedness and minimum maintenance	Safety, lesser running cost

The electric motors can be controlled precisely in a much shorter control period. For multiple in-wheel motors, the individual torque of each motor can be controlled independently. In order to achieve integrated dynamic control of a vehicle, torque of in-wheel motors could be manipulated for incorporation of ABS, TCS, etc. Thus, an electric vehicle with in-wheel hub motors has a high dynamic response due to direct drive to tyres, where advanced motion control techniques can be applied as the motor torque of each wheel and the speed of the vehicle can be controlled precisely. The torque response of an electric motor is in the order of a few milliseconds [25]. The benefits of the use of electric motors as in-wheel powertrain are identified in [26]:

i. In the longitudinal direction, TCS and regenerative braking can be used due to improved torque control

ii. Vehicle handling and stability could be enhanced by the provision of torque vectoring.

But there is certainly an increase in unsprung mass (due to the weight of in-wheel hub motor) and total vehicle weight (due to heavy battery) resulting in a deterioration in the side and a compromise in handling behavior. The independent in-wheel hub motors give efficient powertrain efficiency and also able to provide an improvement of vehicle stability by independent motor torque control of each wheel. The HEVs/EVs plying in the market have used axle motors but most of the studies that investigated vehicle stability control for HEVs/EVs have considered independent braking/traction using hub motors. However, the optimal allocation of the control action to individual wheel brakes and electric motors is yet an open problem. Therefore, control strategies are required to be developed and implemented to enhance vehicle stability and to optimize actuator usage [27].

2.4 Vehicle Dynamics Control (VDC) Systems

The vehicle dynamics control (VDC) systems aim at controlling the vehicle to follow the driver's command and keep the vehicle stable under emergency situations [28], [29]. During normal driving, the driver drives the vehicle when a given steering wheel angle produces a proportional change in the heading of the vehicle. The time lag between the input and the response is very small. But scenario changes when driving an ordinary vehicle at a higher lateral acceleration of above 0.4g on a dry road. The lag time of vehicle response lengthens and as a result, a panic situation arises and the driver loses control [25]. During such scenarios, electronic stability systems augment the driver's inputs and the vehicle is controlled just like in a linear range driving. Thus, lateral stability can be achieved by corrective yaw moment using VDC system. It can be generated by different ways, like by braking individual wheels, by superimposing the steering angle of the driver with a correction (known as active steering), by modifying the steering angles on an axle for multi-wheeled steering system, or by varying the available driving torques on driven wheels (known as torque vectoring control).

Thus, various chassis control systems, such as anti-lock braking, active front steering, traction control VDC system are designed and implemented in vehicles. Numerous studies have confirmed that VDC systems are highly effective in helping the driver in maintaining control of the vehicle, thereby keeping the vehicle on the desired path by leaving the driver fully or partially out of the loop. The VDC enhances driving safety by providing [25]:

i. Improved directional stability under all operating conditions, such as during emergency stops

ii. Increased vehicle stability at the limits of traction, such as during rapid lane change maneuvers

The VDC has been developed to prevent vehicles from spinning and drifting out and often referred to as yaw stability control (YSC) systems. Figure 2.9 schematically shows the function of YSC system. The direct yaw moment control is considered as a suitable way of controlling lateral motion of a vehicle [3]. The purpose of the VDC is to restore the yaw rate of the vehicle as much as possible to the nominal motion expected by the driver. The lower curve shows the trajectory that the vehicle would follow in response to a steering input from the driver if the road had a high tyre-road friction coefficient. The upper curve shows the trajectory that the vehicle will follow if vehicle speed were too high and coefficient of friction is low. The driver may partially succeed in making the vehicle's yaw rate closer to the expected nominal yaw rate, as shown by the middle curve when coefficient adhesion is very small and the vehicle speed is low. Thus, it is clear that VDC systems are needed to be incorporated as the behavior of the vehicle at the limits of adhesion is quite different from its nominal conduct.

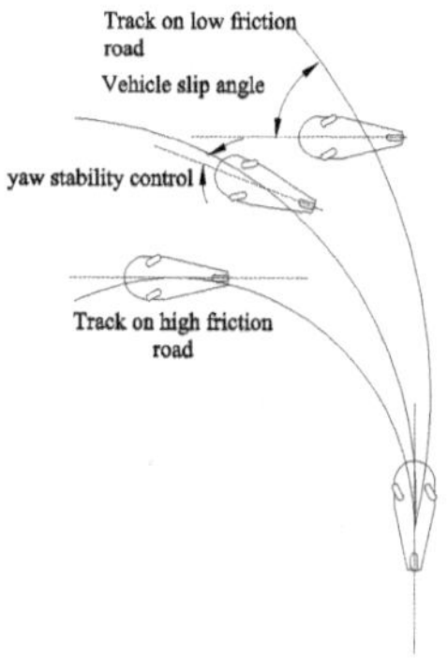

Figure 2.9: The functioning of yaw control system

At large slip angles, changing the steering angle produces very little change in the yaw rate of the vehicle. Studies show that on dry roads, vehicle maneuverability is lost at large slip angles (around 10°), while on packed snow, vehicle maneuverability is lost at low slip angle (around 4°) [30]. The VDC system addresses these issues by reducing the deviation of the vehicle behavior from its normal behavior. The following three types of VDC systems are developed for yaw control [31].

a. Differential Braking system which utilizes anti-lock braking system (ABS) on the vehicle to apply differential braking between the right and left wheels to control yaw moment. This type of system has been implemented on several four-wheeled production vehicles.

b. Steer-by-Wire systems which modify the driver's steering angle input and add a correction steering angle to the wheels for better stability and the research is still hot in this field.

c. Active Torque Distribution systems which utilize active differential and all-wheel drive technology to independently control the drive torque distributed to each wheel and thus provide active control of both traction and yaw moment. It may be noted that torque vectoring can be effective when driving torque exits and when the driver releases the accelerator pedal or applies the brake, intervention by torque vectoring is not possible [32]. [33] presents an extensive review of torque vectoring control (TVC) systems for improvement of the dynamic performance like yaw movement, anti-lock braking and traction controls. Conventionally, equal torques at driven wheels are provided for any vehicle. Thus, constant torque is distributed over driven wheels for the fully electric vehicles. Two different torque vectoring strategies have been mentioned in literature [33], i.e. torque proportional to the wheel vertical load and torque distribution with same longitudinal slip ratio on each wheel. The selection of the most suitable wheel torque distribution for achieving the desired under steer characteristics is required to be optimized.

2.4.1 Integrated vehicle dynamics control system (IVDC)

A passive vehicle chassis functions as an under-steer, a neutral steer or an over-steer depending on the statics and dynamics of the vehicle and the environmental parameters. The dynamic characteristic of the vehicle cannot be altered during dynamic operation. But an active chassis control system of the vehicle has the ability to vary or adapt within a predefined control range based on its dynamic operating conditions. Each individual VDC system like ABS, TCS etc. works in their assigned individual dynamic domains such as longitudinal, lateral and/or vertical dynamics [34]. For example, ABS can influence both longitudinal and lateral dynamics. Thus, the optimum driving dynamics can be achieved when individual VDC systems are networked together in order to control the tyre forces and meet instantaneous driving dynamics safety and comfort requirements. The integrated vehicle dynamic control co-ordinates two or more sub-systems as per control objectives and takes necessary actions with respect to both software and hardware in order to maximize overall vehicle dynamics performances in longitudinal, lateral and vertical directions [35]. The simultaneous use of many control sub-systems poses challenges like more complicated hardware and software due to the increased number of sensors, functional overlapping and conflicting control objectives of different control sub-systems. The challenge is in the process of whole chassis integration of these sub-systems in order to avoid their interventions so as to achieve the overall vehicle dynamics performance. The two key problems of IVDC system which require attention have been identified in [36]:

i. Avoidance of conflicts and interventions among different sub-systems

ii. Use of optional potentials of sub-systems in communicating and coordinating among them since they have different action domains

In the present scenario, the integration technique between steering and braking/traction is being researched and developed worldwide [36]. The more potential benefit of IVDC can be realized once drive-by-wire technology and actuator hardware attain further maturity.

2.4.2 The drive-by-wire system

The drive-by-wire systems use electromechanical actuators and electronic control system replacing the conventional mechanical linkages. Thus, relevant technologies like steer-by-wire, brake-by-wire, throttle-by-wire are being implemented in a vehicle to enhance controller efficiency. It offers benefits such as lower maintenance, better dynamic response and driver assistance functions. Consequently, vehicle manufacturers have launched research programmes towards implementing the full-fledged drive-by-wire technology. An overview of the main features of electric actuators and communication network for drive-by-wire systems is presented in [37]. From a safety point of view, these technologies must provide reliable operation and promptly detect functional degradations through continuous onboard monitoring and hence need to be addressed to ensure operational integrity. For NTTW vehicle, the steering tilt control (STC) scheme requires the use of steer-by-wire technology for its implementation.

2.5 NTTW Vehicle Tilting and Control

As compared to a standard vehicle, the track width of a narrow vehicle is much less resulting in an increase in ratio of center of gravity to track width. This poses stability problem for narrow vehicle while cornering. Hence, there is a need to lean the vehicle body towards inside while cornering.

The desired tilt angle (θ_{des}) for tilt equilibrium is the angle achieved by balancing the centrifugal force during cornering with the weight of the vehicle and is calculated by [6]:

$$\theta_{des} = \arctan \frac{a_y}{g} = \arctan \frac{\dot{\psi} V_x}{g} = \arctan \frac{{V_x}^2}{g R_c} \tag{2.1}$$

It is understood from the equation (2.1) that if the vehicle velocity is increased while negotiating a turn at a fixed tilt angle (fixed lateral acceleration), the radius of curvature is required to be increased. This behavior reflects the characteristics of an understeer vehicle [14]. It can be also observed that the steady-state front wheel steering angle is a function of the yaw rate ($\dot{\psi} = V/R$) and the tilt angle of the vehicle body. Hence, a unique steady-state steering input (δ_{ss}) is required for a steady state tilt angle (θ_{ss}) under a certain yaw rate (fixed V and R). Thus, for an NTTW vehicle tilt dynamics is added to the lateral dynamics for analysis of vehicle dynamics during cornering of the vehicle.

2.5.1 Tilt dynamics

Narrow tall vehicles tend to tilt and overturn. The driver of a narrow vehicle such as a motorcycle is normally responsible for balancing the vehicle. In addition, while negotiating a curve, the driver must tilt the vehicle into the curve so as to compensate for the tilting moment of the centripetal force generated by tires. The use of an active tilt control system to assist the driver in balancing as well as automatic tilting while cornering is expected to greatly enhance the acceptability of narrow vehicles [1], [38], [39]. In addition to the longitudinal, lateral and vertical dynamics, provision of strategy for additional active tilt stability dynamics is essential in narrow three wheel platform. The interdependencies amongst longitudinal, vertical, lateral and tilt dynamics are explained in Figure 2.10. The longitudinal, lateral and yaw motions of a vehicle are primarily determined by tyre friction forces at the contact patch between tyre and road. The tyre friction forces have limits, depending on road conditions and the vertical tyre force (F_z), which is represented by tyre friction ellipse. Thus, the optimum driving dynamics can only be achieved when tyre forces on all the wheels and in all directions can be influenced and controlled precisely. The friction eclipse couples the lateral and longitudinal tyre forces and limit to the longitudinal friction coefficient with respect to longitudinal slip. This limit is dependent on longitudinal and vertical tyre forces, whereas the limit to the lateral friction coefficient with respect to lateral slip is dependent on lateral and vertical forces of tyres.

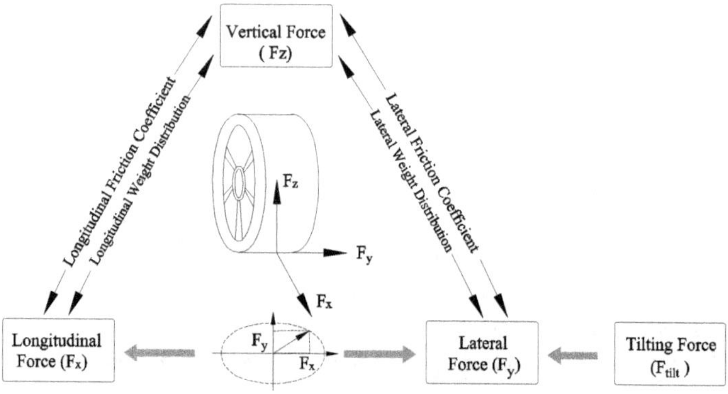

Figure 2.10: Interdependencies between the degrees of freedom of NTTW vehicle

For a tilting vehicle, the force needed for achieving vehicle body leaning to the desired angle is added to the lateral force. However, [15] pointed out that the equilibrium as required for the stability of NTTW vehicle is not fixed as velocity varies with time and the desired tilt angle changes as velocity is changing. This makes the driving stability analysis more difficult, which will be discussed later. The leaning of the vehicle body at a desired angle (which is continually changing) demands incorporation of an active control system to ensure the tilt stability without affecting the handling behavior of the vehicle. The stability requirement for the lateral dynamics (trajectory tracking) and the tilt dynamic cannot be simply superimposed in order to stabilize a narrow tilting vehicle. This is noted in [40]. This is due to the fact that they demand opposing steering inputs while negotiating a curve in order to generate lateral forces that are directed in the direction and in the opposite direction of the center of rotation.

The dynamic maneuvers (like a figure of 8, rapid lane change) scenarios for any vehicle require instantaneous changes in steering input with high steering rates. These pose the problem with control strategies to put in practice. The NTTW vehicle body leans to the side during cornering in order to balance the moment caused by the vehicle lateral acceleration and the moment caused by the gravity. The control objective is to continuously ensure a balance between the forces from lateral acceleration and the gravitational pull. This leads to the question of whether an optimum tilting profile exists wherein the motion of the vehicle body and the build-up of lateral acceleration are synchronized in such a way that the steady state demand can be reached as rapidly as possible without the vehicle rolling over. A study was conducted by [6] by assuming a profile for the lateral acceleration and tilt acceleration and optimizing it to maximize the lateral velocity of the vehicle without roll over. Authors concluded that independent control of the tilting and lateral acceleration is required in order to optimize the lateral dynamics. Thus, the direct link between the driver input at the steering wheel and the front wheel steering angle is necessary to be broken. In other words, SDTC strategy is required to maximize vehicle maneuverability without exceeding the limits of stability.

2.5.2 Tilting mechanism for leaning of vehicle body

The mechanism to tilt the vehicle body can be of two distinct types:

i. **Full vehicle body tilting**:
   The tilting of vehicle body using leaning of camber has been demonstrated by many researchers [39], [41] and [17]. The concept of the lean mechanism is explained in Figure 2.11. The leaning mechanism could be considered as an extension of four bar linkage. A tilting actuator in the form of rotary actuator is used for controlled rotation of parallel arms which determines the lateral inclination of the wheels and therefore the roll of the vehicle body. Whenever one wheel goes over a bump or ditch, the mechanism works as conventional wishbone suspension systems.

Figure 2.11: Schematic of full body tilting of NTTW vehicle

ii. **Partial vehicle body tilting**:

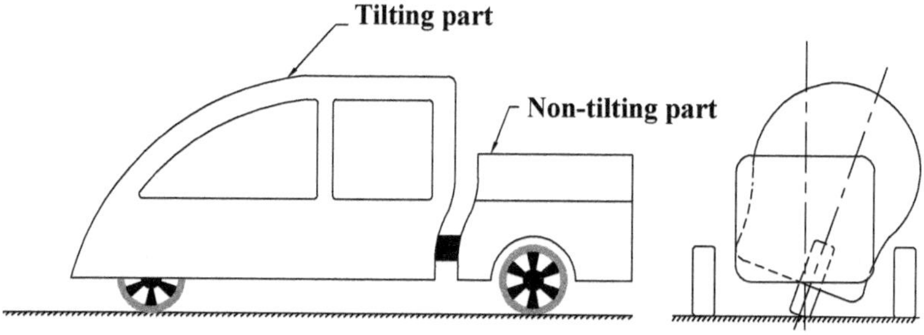

Figure 2.12: Schematic of a partial body tilting of NTTW vehicle

A significant amount of study has been carried out in Bath University, UK on a configuration of NTTW vehicle with leaning front portion of the vehicle body and non-tilting rear part [6] [19]. The rear part houses the engine and has two wheels. The front part accommodates the driver and leans to a side while negotiating a turn. The concept is shown in Figure 2.12. The heavy rear part is non-tilting and hence there seems to be some advantage for the design of tilting control strategy as tilt joint stiffness plays a major role in the dynamics response.

2.5.3 Tilt control system of NTTW vehicle

In a majority of the literature [42], [39], [43] the emphasis has been given to the combination of the lateral and tilting dynamics while cornering maneuvers. Longitudinal dynamics is not included in most of the simulation models and have considered constant longitudinal velocity [44]. However, [45] and [46] have considered longitudinal speed as one of the states of the vehicle model. The authors in [45] have designed tilt angle and speed controllers using sliding mode framework. Authors in [46] have presented a rudimentary trajectory generator and PI/PD strategies for controlling lateral, longitudinal speeds and the tilting reference. The authors in [4] proposed a method to find optimal tilt angle for various speeds. [5] mentions that one of the fundamental problems of three wheeled vehicles is their stability under braking or acceleration. [47] has implemented gain scheduling technique to account for variation in longitudinal velocity. The vehicle body can be leaned by means of two mechanical systems. Firstly, by means of a tilt actuator which provides a tilting torque at the center of gravity, known as direct tilt control (DTC) system. Secondly, tilting by employing counter-steering effect using steer-by-wire assembly, which is called steering tilt control (STC) system. DTC is generally effective at lower speeds and STC at higher speeds. Hence a combination of the two is the desired tilt control scheme having advantages of both systems. It is called steering direct tilt control (SDTC) system [14], [48]. All the three types of tilting control schemes are explained in more detail below.

(a) Steering tilt control (STC) system

The STC scheme takes driver's input and appropriately scales it to determine the desired tilt angle [14]. A simple STC based on proportional control strategy is proposed in [49]. No special skill of the driver is required for STC and as such driver's input is not directly linked to the steering mechanism. For STC, the mechanical link between the driver's steering wheel and the steered wheels is broken and a form of active steering system is introduced. The countersteering is used by single track vehicle (cycle, motorcycle) riders to initiate a turn and the required lean in the direction of that turn by momentarily steering in the opposite direction (steer right to turn left). The lateral force developed at tyre-road contact path creates a torque around the longitudinal (roll) axis of the vehicle and this torque causes the vehicle to lean away from initially steered direction and towards the direction of the desired turn. This scheme is very effective at high speed by providing a balancing torque by the way of steering to right initially [8] and then to left for making a turn to the left. One disadvantage of STC is that it does not work at all at the stationary condition and requires higher gains for stability for stability as speed decreases [42]. Also balancing becomes unsafe due to the lack of tyre-road adhesion in slippery road conditions with this type of control strategy. The authors in [40] have proposed an automatic STC system based on PD control strategy. This steer-by-wire design utilizes the driver's output as the desired tilt angle. The steering angle applied by the driver is manipulated by the STC steer-by-wire system by providing by providing a counter-steer angle so as to obtain correct steering angle for STC strategy. Also, the STC method has the underlying issue of underactuation, as it has to control two actions,i.e. steering and tilting, with only one input, i.e. the driver steering angle. Research has been done for the design of control strategies for STC systems. [14] and [41] have proposed three different methods to calculate desired signals to design STC using PD controller. The STC controller determines the desired steering angle to achieve the desired tilt angle which is calculated using the driver's steering and speed commands to maintain stability (the schematic of the control strategy is shown in Figure 2.13). A new strategy employing PID control technique has been developed for STC to work in synchronism with driver's model in [2]. The aim of the study was to minimize the deviation of the vehicle trajectory from the desired path. A sliding mode controller if proposed for STC along with vehicle longitudinal velocity in [45]. STC system is not suitable for low longitudinal velocities, for e.g.less than 8 ms^{-1} [42] as large counter-steering is required to lean the vehicle. This scheme is very effective at high speed by providing a balancing torque by counter-steering, which if high, essentially deviates the vehicle from the desired trajectory. However, [45] has tried to resolve this issue by incorporating trajectory tracking. The trajectory tracking still needs to improve, as the tracking of tilt angle takes precedence and therefore, steady state-error is observed in speed tracking.

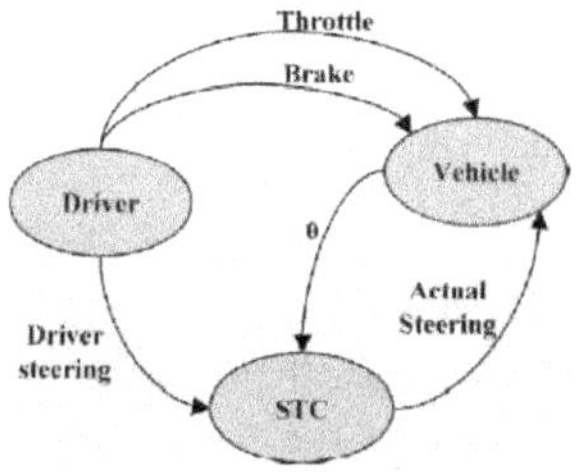

Figure 2.13: Schematic diagram of vehicle control system based on STC system [14]

(b) Direct tilt control (DTC) system

In the DTC scheme, an actuator is used to lean the vehicle body and to controls the tilt stability of the vehicle. The actuator tilts the vehicle body during cornering at a desired angle based on the radius of curvature and the velocity of the vehicle. The DTC is found to be beneficial during low-speed driving. The disadvantage of DTC is the delay in vehicle response and the risk of vehicle oscillations and consequently, it requires complex control strategies suitable to different loading and driving conditions. Another inconvenience is the high power requirements to generate the required tilting motion, as the tilting motion always acts against the centrifugal forces. The steering torque can be used as input parameter for quicker vehicle response and makes it possible to tilt the vehicle even before the cornering has begun.

The scenarios investigated in the works of [6] have been confined to constant forward velocity situations. The objective was to find the optimum tilting profile that would maximize the lateral velocity of the vehicle without causing roll-over. The author proposed a control system that treats the driver steering input as a lateral acceleration demand that is to be reached as rapidly as possible and with minimum load transfer across the rear axle. Similarly, several control strategies are devised to control the active tilting of the NTTW vehicle using DTC mechanism. Two different controllers are designed and analyzed in [44] to provide the desired tilt angle during cornering maneuver. A receding horizontal controller which previews apriori information on road curvature has been used in conjunction with two baseline stabilizing controllers based on state feedback law so as to minimize the tilting torque requirements. The second approach applies a nonlinear control method which uses feedback linearization and is found to be more effective in minimizing required tilting torque. A PD controller for DTC system by employing different methods to calculate desired tilt angles are designed and presented in [14] and [41]. The main aim was to find a controller so that the required transient tilting torque is minimal and zero steady state torque is achieved. The overall flowchart of the DTC system is shown in Figure 2.14.

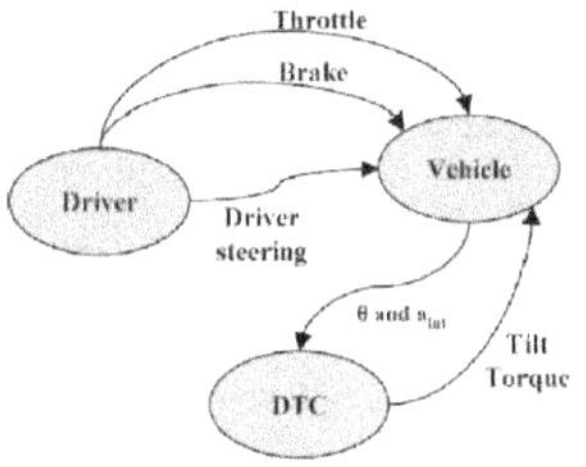

Figure 2.14: Schematic of vehicle control system based on DTC system [14]

A control strategy for DTC system based on LQ controller to directly achieve zero perceived lateral acceleration without using intermediate values of the tilt angle is proposed in [50]. This, along with the usage of a feed-forward function of steering angle, reduces the required maximal tilting torque. The control strategy discussed earlier [14] and [41] is based on achieving tilt angle equal to the desired tilt angle (θ_{des}). This strategy has limitations due to approximation in obtaining θ_{des} and hence, leading to non-zero steady state torque in a DTC control strategy. Also, the perceived acceleration is not really zero. This control strategy is sensitive to the road bank angle. A control strategy based on controlling directly the lateral perceived acceleration on the vehicle body (a_{per}), without using the intermediate computation of the tilt angle (θ_{des}) is proposed in [51]. Here a zero steady state torque with zero perceived acceleration is achieved simultaneously at a constant curve path. The authors have further developed a gain-scheduled controller to obtain the validity of the proposed controller over a speed range. [6] have proposed a DTC system for hydraulics control employing PID strategy. The DTC scheme utilizes the steering input to calculate the desired tilt angle which makes the control design comparatively simpler. But, the requirement of high tilting torque by the tilting actuator at higher speeds is a disadvantage. Also, if the tilting and steering are not synchronized, the transient torque increases which creates a discomfort for the driver.

(c) Steer direct tilt control (SDTC) system

It is logical to use DTC system at low speeds and STC system at high speeds [42]. The SDTC is a combination of STC and DTC systems. In this type of control, the proportionment of STC and DTC controllers are required so that at lower speeds DTC system is dominant and at higher speeds, STC system comes into action, which is implemented by using weighting functions. When both STC and DTC act simultaneously in parallel, both the controllers should exert control effort in the same direction. But as understood from the basic operating principle of both type of control systems that STC and DTC will try to tilt the vehicle body in opposite direction during the transient phase. When DTC is used at low speeds and STC at high speeds, then a switchover speed is needed, where switch over from one type of control strategy to another is required to be performed. [14] showed that the start and end of the switchover speeds are at 5 and 10 m/s velocities and inferred that DTC system, where θ_{des} is calculated based on lateral acceleration on the vehicle body during cornering, is best suited for efficient utilization of actuator efforts. Thus, an integrated DTC and STC (i.e. SDTC) system can be used for tilt stability of a tilting vehicle, where θ_{des} should be calculated based on driver steering input since it ensures conflict free operation with STC system. The Figure 2.15 shows the schematic of the overall concept of the control strategy for SDTC system used in [14]. The subject control problem can be solved by the use of the driver's steering input to generate the desired tilt angle of the vehicle where control law follows a scaling factor to rotate the desired tilt angle with the driver's steering input. This ensures that both the STC and DTC systems exert control efforts to tilt the vehicle to the same tilt angle at all times. However, the transient torque needed for DTC system will not be optimized. Also, experimentations showed

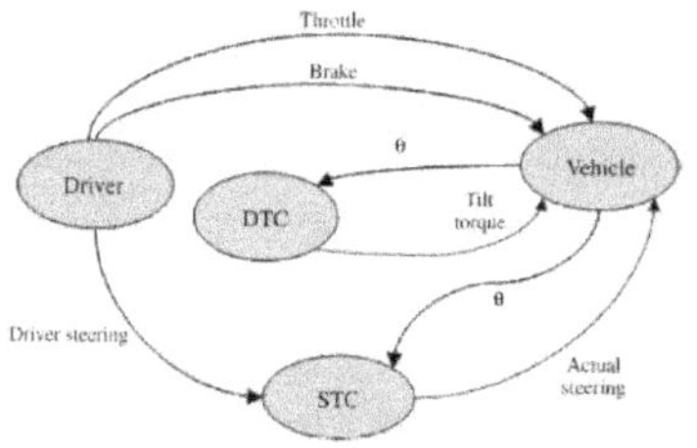

Figure 2.15: Schematic of vehicle control system based on SDTC system [14]

that a very good desired tilt angle tracking was achieved up to 10 degrees of tilt angle.

As pointed out earlier, the control laws are required to be designed accordingly and two systems (i.e. STC and DTC) have different control signals and these signals react totally different ways during transitions. [42] experimented with the use of a threshold to shift from one control system to another at a preselected speed (between 7.5 m/s to and 8.5 m/s) through computer simulations. The merit of each system was evaluated through simulations and it was observed that switching from DTC to STC or vice versa for nominal vehicle model was satisfactory. [6] also advocated for use of SDTC based on optimization study made to maximize the lateral velocity of the vehicle without roll-over. An SDTC system where both STC and DTC systems work simultaneously to achieve a common tilt angle demand during various speeds for the CLEVER vehicle is proposed in this paper. The controller used the steer demand of the driver and the speed of the vehicle to estimate lateral acceleration and therefore, the required tilt angle in order to stabilize the vehicle. Experimental roll stability performance verification of the SDTC system under severe maneuvers is proposed in [19]. It is observed that the wheel load variations in SDTC system were 36 % lower than that of DTC system at 10 m/s forward speed for CLEVER vehicle.

The authors in [41] have proposed an SDTC control, wherein switching between DTC and STC is obtained as a function of vehicle longitudinal speed (control system schematic is shown in Figure 2.15). An SDTC strategy to achieve a common tilt angle (for both STC and DTC systems) demand throughout the entire speed range of the vehicle is proposed in [19] and [48]. The controller uses the steer demand of the driver and the speed of the vehicle to estimate the required tilt angle in order to stabilize the vehicle. Additionally, a modified front steer angle is created to compensate for the trajectory tracking error. The overall block diagram of this strategy is shown in Figure 2.16. [6] concluded that independent control of the tilting and lateral acceleration is required to optimize the lateral dynamics. To achieve this, the link between direct input at the steering wheel and corresponding front wheel steering angle must be removed.

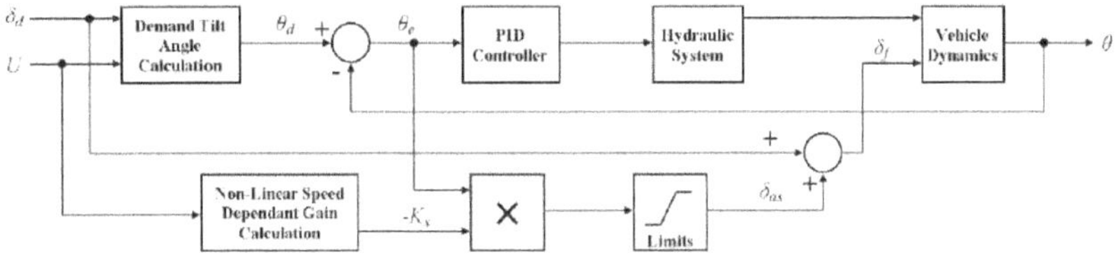

Figure 2.16: Schematic of vehicle control system based on SDTC system [19]

A multivariable approach based on H_2 control strategy, wherein both STC and DTC systems function in a systematic combination is considered in [51] and [47]. The authors have mentioned that no separate driver controller is designed as it was a complex task. Since at low speed, the lateral acceleration is very low, the vehicle can be locked in the upright position. Hence at θ_{des} equal to zero, the tilt torque requirement is also zero. The use of tilt lock using tilt brake system has been tried out in a prototype by [14]. When the tilt lock is engaged, the driver's steering input is directly passed on to the steered wheels. Also, low-speed maneuvers like parking become easier for the driver. However, this tilt lock brake can function above a defined velocity of the vehicle. Once the tilt brake is disengaged, the steering will be governed by the STC control law. The tilt brake system ensures a smooth ride for the driver while avoiding overshoots in tilt torque and steering input requirements by engaging/disengaging the tilt brake system as per need. However, no further work on the use of tilt brake has been reported.

2.6 HIL setups in automotive technology

The hardware-in-the-loop (HIL) simulation is a technique that is used during the development and testing of complex control system [52], [53]. The complexity of the plant under control is accounted for by adding a mathematical representation (plant simulation) of all related dynamic systems. The device-under-test interacts with this plant simulation. In a typical HIL simulation, a real-time processor executes mathematical model emulating system dynamics. In the context of automotive applications, the HIL simulation set-up is an efficient tool that can save cost and development time towards system verification. HIL is used for performing various system-level tests. For automobiles, they could be the prime mover, brake, steering and other systems. In the literature, several HIL setups on specific applications are reported. For example, HEV/EV drive train components can be tested on the HIL simulator during the design process so that the real-time performance can be evaluated [54], [55], [56], [57], [58]. A more practical environment was emulated by the inclusion of compressed air system with associated anti-skid brake-line components in the test setup [52]. The authors have used less expensive xPC target enabling real-time simulation on a PC platform. The vehicle, the tyre and anti-skid brake system (ABS) models are developed in Matlab/Simulink in the host PC. The ABS controller algorithm has been validated using HIL setup. The effects of different control parameters on the vehicle-driver-controller interaction in different vehicle maneuvers for novice and expert drivers are compared in [59]. The authors have utilized a virtual test track consisting of real-time simulation hardware, a visual graphics engine and input/output systems (Figure 2.17).

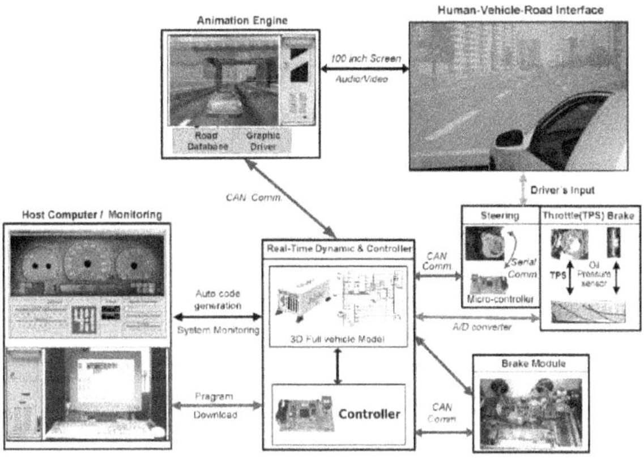

Figure 2.17: Configuration of the hardware-in-the-loop set up used in [59]

The yaw stability controller in a real-time HIL simulator with the driver in the loop (shown in Figure 10.2) is validated in [60]. The vehicle model worked on a dSpace (DS 1103) simulator system and the controller was implemented on xPC target box. The communication between them was carried out using the controller area network (CAN) bus. The driver inputs are connected to the dSpace unit through its analog and digital channels that directly connected them to the vehicle model. The visualization for the driver is handled using the dSpace motion desk software.

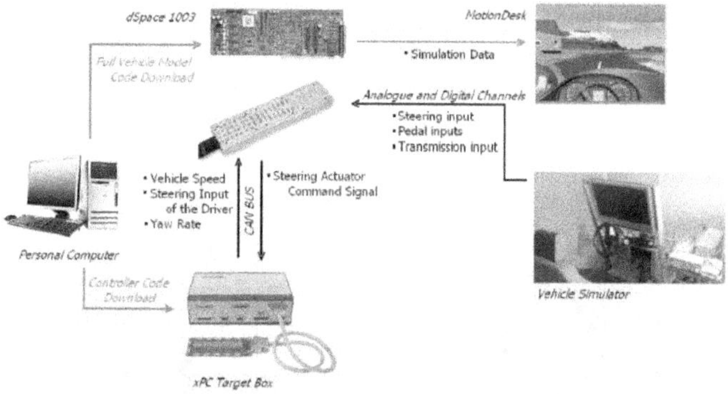

Figure 2.18: Configuration of the hardware-in-the-loop set up used in [60]

2.7 Summary

i. Majority of research work, as observed from literature survey, is centered on modeling of NTTW vehicles to investigate various control strategies for roll-over prevention and tilting during cornering. The issue of coupling of tilt dynamics with other aspects of vehicle motions such as longitudinal dynamics, dynamic vertical tyre forces are not fully addressed. An integrated control system to provide dynamic stability for three wheeled tilting vehicle system under dynamic conditions is also not found in the literature. Additionally, there is a scope to enhance the stability of NTTW vehicle by utilizing electric in-wheel motor as the drive power.

ii. The literature survey on tilt control systems for NTTW vehicle indicate that the SDTC system or the combination of STC and DTC systems provides a better dynamic stability control solution.

iii. Overall control objectives need to be evolved to include the coupling amongst the dynamics to regularize the objectives studied and analyzed by the past researchers.

iv. The crucial parameters are needed to be chosen keeping practicability of NTTW vehicle specifications in view to model and analyze dynamic stability control strategies to realize a realistic controller design.

Chapter 3

Mathematical modeling of NTTW vehicle and its parametrization

3.1 Introduction

It is necessary to analyze the dynamic behavior of NTTW vehicle platform before initiating the design of stability controller. This requires derivation of a mathematical model that describes the system properties to acceptable accuracy [61]. The basic mathematical model based on Ackermann theory relies on only geometrical parameters of the vehicle and tyre characteristics are not included. The geometric equations overestimate the states and the resultant model is found to be inadequate for the development of any stability control system [2]. One of the initial mathematical models for NTTW vehicle was proposed by [42] which was based on kinematic linking of steering with the tilting body which addressed static plant behavior. But such models only partially describe transient behavior of dynamic performance of the vehicle. A five DoF multi-body lateral dynamics model of the NTTW vehicle is presented in [6]. The author has also included the dynamics of the hydraulic actuators which tilts the rear body of the vehicle in the overall modeling. The same model is used for simulation studies by [19]. A nine degrees-of-freedom tricycle model using Lagrange's equation is formulated by [45] which is designed for vehicle equipped with steering and traction motor actuators modeled with first order transfer function. Authors in [16] and [15] have proposed a novel dynamic model based on switching between multiple running states of the NTTW vehicle, such as 'normal running' or 'temporary running' while right wheel not on the ground. This approach provides a simulation platform to test risky driving conditions. Significant amount of studies in the modeling of the NTTW vehicle are performed by the University of Minnesota [39], [44], [14], [41]. Several linear and non-linear lateral dynamics models using first principle method have been proposed. A simple three DoF bicycle lateral dynamics model is utilized by [47] for control design and is extended by [46] by including longitudinal speed dynamics.

This chapter presents the development of dynamic modeling of the NTTW vehicle used in this research work. The study involves the development of a mathematical model of the NTTW vehicle which encapsulates the relevant characteristics of the vehicle. This includes the typical forces acting at tyre-road interaction points (tyre model) and also other dynamic forces (such as vertical tyre forces) encountered during various maneuvers of such a vehicle. This mathematical model is intended both for understanding the behavior of an NTTW vehicle during dynamic maneuvering and to further assist in the design of active tilt control systems. The response of individual sub-systems of NTTW vehicle with basic open loop tests is also presented in this chapter.

3.2 Reference frame

The system of coordinates reported in this present work is as shown in Figure 3.1. It is in accordance with the ISO standard 8855.

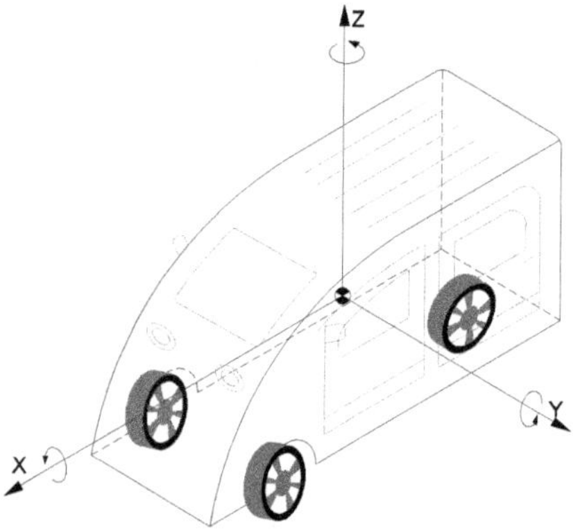

Figure 3.1: Reference frame used for modeling of *NTTW* vehicle body

In this reference coordinate system, the forward movement of the vehicle is described in the positive X axis, the lateral movement is described by the Y axis, being positive when oriented to the left (from the driver position) and the vertical movement is represented in the Z axis. The roll rotation (tilting of vehicle body) is defined around the X axis, the pitch rotation around the Y axis and the yaw rotation around the Z axis.

3.3 Non-linear Mathematical Model of Proposed NTTW Vehicle

It can be observed from the literature associated with modeling of vehicle dynamics that there are numerous degrees of freedoms associated with vehicle dynamics. But the choice of the variables to consider in any research depends on the aim of the study. As grounds of good research work, it is appropriate to first decide on the assumptions that are accounted for the formulation of the NTTW vehicle model:

i. The entire mass of the vehicle is considered to be at the center of gravity (CG)

ii. The vehicle body tilt angle is assumed to be equal to the wheel tilt angle,i.e. wheels tilt along with the body

iii. The vehicle is moving on a leveled road, i.e there is no banking of the road

iv. The suspension, roll and pitch dynamics, as well as gyroscopic effects due to rotation of wheels, are not considered

In this study, a detailed non-linear mathematical model of an NTTW vehicle chassis dynamics along longitudinal, lateral and tilt (of body) motions is implemented. The NTTW vehicle under consideration is assumed to have two wheels in the front and one in the rear. In order to design chassis stability control system for a vehicle, it is pertinent to know the dynamic forces experienced by the vehicle due to driver maneuvers and also due to the road conditions. Hence, the simulation model with capture these details with maximum fidelity as possible. The global axis is represented by XYZ as given in Figure 3.2. The local axis xyz is attached to the CG of the vehicle with xy as the horizontal plane. The yaw angle (ψ) is measured with respect to the global axis X. Front and rear lateral tyre forces in XY plane are represented by F_{y_f} and F_{y_r} respectively. There is a need to use another axis $x'y'z'$ which is attached to the body of the vehicle at the CG and which moves with the leaning vehicle. The angle between the upright position of the vehicle body and its leaned position is termed as tilting angle (θ). Then, perceived acceleration (a_{per}) acts along y' axis whereas lateral acceleration (a_y) acts along the y-axis. The explanation of a_{per} is given in Chapter 4.

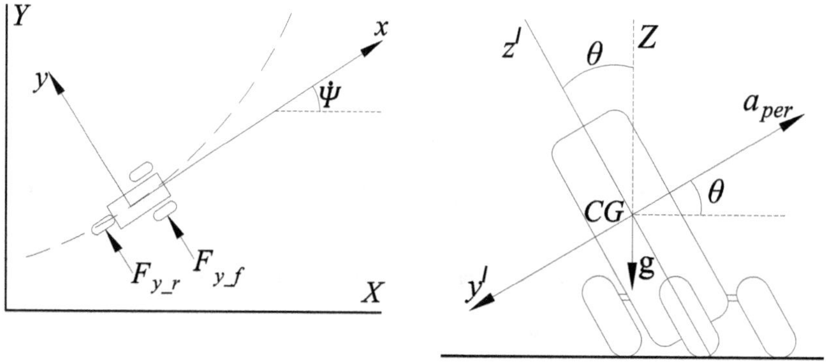

Figure 3.2: Three degrees of freedom of NTTW vehicle body

The mathematical model is derived from first principle method and it can be sub-divided into five major subsystems. Some of the subsystems represent the vehicle motion in a certain degree of freedom and other depict the peripherals. The body dynamics includes the lateral, tilt and longitudinal states of the NTTW vehicle. The peripherals consist of the vertical tyre loadings and the tyre-road interaction forces in lateral and longitudinal forces. These sub-systems are coupled dynamically and hence interact during the vehicle maneuvers. The inter-relationship within the subsystems render the dynamics of the vehicle as highly non-linear and closer to practical behavior. The system block diagram of the vehicle dynamics model is shown in Figure 3.3. In the subsequent sections, NTTW vehicle body model, tyre model, and vertical dynamics model are explained. All these systems interact and simulate vehicle movements synergetically as per driver's commands. The notations used in the mathematical model are given in Table 3.1.

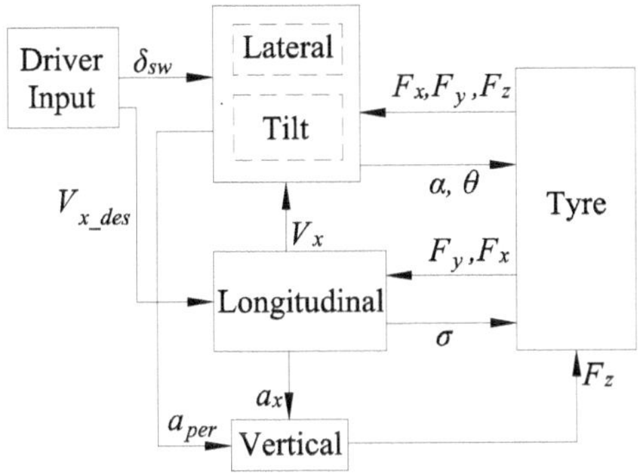

Figure 3.3: Block diagram of NTTW vehicle dynamics model

3.4 Tyre model

The interaction of tyre with the vehicle platform is one of the most important aspects of a vehicle chassis model. The tyre connects the chassis to the road and enables the vehicle to maneuver as per requirement by generating the relevant tyre forces. Modeling a tyre is a complex phenomenon. Many factors influence the tyre characteristics, like the variation of stiffness with inflation pressure, temperature, load. Tyre manufacturer's data, which are not available easily is needed for modeling purpose. Typically, linear tyre model has been used by researchers to calculate lateral forces, whereas [6] has used Pacejka tyre model. Pacejka tyre model is generated by fitting experimental data as pre-requisite. But acquiring such data is time-consuming and expensive and hence it is generally not preferred. The authors in [45] have utilized linearized Pacejka formulation by considering constant stiffness coefficient. In this thesis, an analytical Dugoff tyre model equations are [31] used to calculate lateral and longitudinal tyre forces. The variables used in the tyre model are tyre slip angle, tyre slip ratio and road friction coefficient. The tyre stiffness coefficients are calculated based on certain assumptions as explained in Appendix A and are listed in Table 3.3.

Table 3.1: Notations used for NTTW vehicle model

Notation	Parameter	Unit
V_y	Lateral velocity	ms^{-1}
V_x	Longitudinal velocity	ms^{-1}
θ	Tilt angle	degree
$\dot{\theta}$	Rate of change of tilt angle	$rads^{-1}$
$\dot{\psi}$	Yaw rate	$rads^{-1}$
ω_i	Angular velocity of ith wheel	$rads^{-1}$
F_{x_i}	Longitudinal tyre forces of ith wheel	N
F_{y_i}	Lateral tyre forces of ith wheel	N
δ_f	Front wheel steering angle	$radian$
δ_{SW}	Steering wheel angle	$radian$
T_{tilt}	Moment applied at CG of the NTTW vehicle to lean it	Nm
T_{d_i}	Drive torques of ith wheel	Nm
T_{b_i}	Brake torques of ith wheel	Nm
F_{aero}	Aerodynamic drag force	N
F_{roll}	Rolling drag tyre force	N
F_{z_i}	Tyre vertical force	N
σ_i	Slip ratio of ith wheel	-
α_i	Slip angle of ith wheel	radian
d_{tyre}	Diameter of tyre	m
a_{y_ro}	Threshold rollover acceleration	ms^{-2}

3.4.1 Tyre slip angle

Tyre slip angle is one of the important parameters which is responsible to generate lateral tyre force, also known as cornering force. This force is responsible to make the vehicle turn which means, more the lateral force, less is the driver's need to slow down the vehicle while taking corners. The slip angle of a tyre is defined as the angle between the orientation of the tyre and the orientation of the velocity vector of the wheel [31]. Vehicle dynamics analysis includes relationships between slip angles at the front and rear tyres, and vehicle performance output variables, such as yaw rate and lateral vehicle velocity [62]. A schematic layout for NTTW vehicle showing tyre slip angle while turning is shown in Figure 3.4.

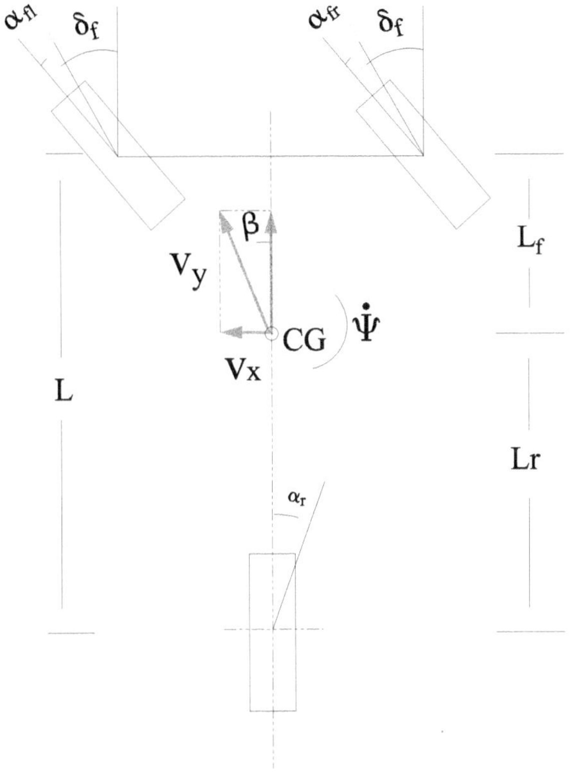

Figure 3.4: Definition of tyre slip angle

The tyre slip angles of front left, front right and rear wheels for NTTW vehicle are represented by equations (3.1) through (3.3):

$$\alpha_{fl} = \delta_i - \frac{V_y + L_f\dot{\psi}}{V_x - \frac{L_{tw}}{2}\dot{\psi}} \tag{3.1}$$

$$\alpha_{fr} = \delta_i - \frac{V_y + L_f\dot{\psi}}{V_x + \frac{L_{tw}}{2}\dot{\psi}} \tag{3.2}$$

$$\alpha_r = -\frac{V_y - L_r\dot{\psi}}{V_x} \tag{3.3}$$

Conventionally, for four wheelers the cornering force is generated through slip angle. However, for two wheelers, the cornering is generated using camber thrust by leaning the vehicle in corners. For the NTTW platform, the cornering force is generated by augmenting the effect of camber thrust with the slip angle, which has been incorporated during formulation of Dugoff tyre model as derived in equation (3.8).

3.4.2 Tyre slip ratio

The longitudinal slip is defined by the difference between the speed of the wheel at the contact patch and the actual vehicle speed. Slip ratio is a measure of tire slipping with respect to the road along the longitudinal direction [63]. When the vehicle accelerates the driven wheels (tyres) exerting a torque on the road, the road pushes the tyre forward in reaction. During this phenomena, the tyre, in reality, slips on the road. It signifies that the tyre rotates faster than the actual speed at which the vehicle is moving. The tyre longitudinal force increases with the increase in tyre slip till the wheel spin occurs (unstable condition). For the NTTW vehicle used in this research work, the longitudinal forces are generated by the front two driven wheels. The rear wheel being non-driven, does not generate the slip and hence no thrust. Longitudinal slip definitions are different in case of acceleration and braking. The wheel slip ratio is represented by equation (3.4) and 3.5):

$$\sigma_i = \frac{\mid \omega\_i(\frac{d_{tyre}}{2}) - V_x \mid}{\omega\_i(\frac{d_{tyre}}{2})} \qquad (acceleration) \tag{3.4}$$

$$\sigma_i = \frac{\mid \omega\_i(\frac{d_{tyre}}{2}) - V_x \mid}{V_x} \qquad (braking) \tag{3.5}$$

3.4.3 Road friction coefficient

The interaction between tyre and road is crucial to vehicle dynamics and control as the tyre forces experienced from the road significantly affect dynamic behavior of the vehicle. Tyre-road friction coefficient (μ) determines the tyre forces that can be generated for a vertical load on the tyre. The value of μ varies between zero and one depending on the type of road surface under consideration, such as icy, snow-covered, gravel, and dry asphalt. In this thesis, μ is derived by standard Pacejka magic formula which converts slip ratio (σ_i) to friction coefficient (μ). The formula is represented by equation (3.6) [64].

$$\mu = a(1 - e^{-b\sigma} - c\sigma) \tag{3.6}$$

Here, the a, b and c are coefficients which are dependent on the road surface that vehicle is being driven on. In this work, it is assumed that road is dry asphalt and hence the values of the coefficients are a = 1.28, b = 23.99 and c = 0.52 [64].

3.4.4 Formulation of the Dugoff tyre model

The Dugoff model assumes a uniform vertical pressure distribution on the tyre contact patch. Additional lateral force can be produced from the change in the camber angle (camber thrust) and thus total lateral force produced is dependent on tyre cornering stiffness (C_α) and the camber stiffness (C_γ) [4]. The longitudinal tyre force (F_x) and lateral tyre force (F_y) are represented by (3.7) and (3.8) respectively. The tyre road friction coefficient (μ) used in equation (3.8) has been deduced from equation (3.6).

$$F_x = C_\sigma(\frac{\sigma_i}{1+\sigma_i})f(k_t) \tag{3.7}$$

$$\begin{aligned} F_y &= C_\alpha(\frac{\tan\alpha_i}{1+\sigma_i})f(k_t) + C_\lambda\theta \qquad (3.8) \\ where, & \\ f(k_t) &= (2-k_t)k_t ....if \quad k_t < 1 \\ f(k_t) &= 1..........if \quad k_t \geq 1 \\ and & \\ k_t &= \frac{\mu F_z(1+\sigma_i)}{2\sqrt{(C_\sigma\sigma_i)^2 + (C_\alpha\tan\alpha_i)^2}} \end{aligned}$$

The functionality of tyre model is explained as follows. The tyre model requires inputs viz., slip angle, slip ratio and the body tilt angle, from the lateral, longitudinal and tilt dynamics models respectively to calculate the tyre forces. The forces generated from the tyre model are input to the longitudinal, lateral and tilt dynamics models in equations (3.17) to (3.23). Here, σ_i denotes the slip ratio where 'i' can be fl (front left) or fr (front right). The d_{tyre} and α_i denote the diameter and the slip angle of tyre respectively.

3.4.5 Tyre characteristics

The characteristics of the tyre used in this work are shown in Figure 3.6 and 3.5. The tyre characteristics can be graphically represented by two major tyre forces, i.e. lateral and longitudinal. Figure 3.5(a) shows the plot of lateral tyre force against slip angle. Whereas, the Figure 3.5(b) represents the plot of longitudinal tyre force as a function of slip ratio wherein it is implied that the maximum longitudinal force for the chosen road condition is generated at a slip ratio of about 30%. The relationship of F_y Vs F_x for varying slip ratios or slip angles is shown in Figure 3.6. It is observed from simulated results that for a constant slip angle, as the slip ratio increases, the limiting envelope of the tyre forces increases (shown in Figure 3.6(a)). Similarly, for a constant slip ratio, as slip angle increases, the limiting envelope of tyre forces also increases, as shown in Figure 3.6(b).

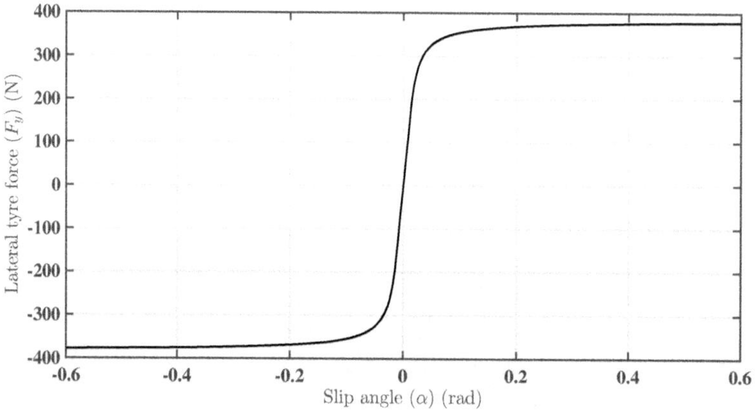

(a) Lateral tyre characteristics

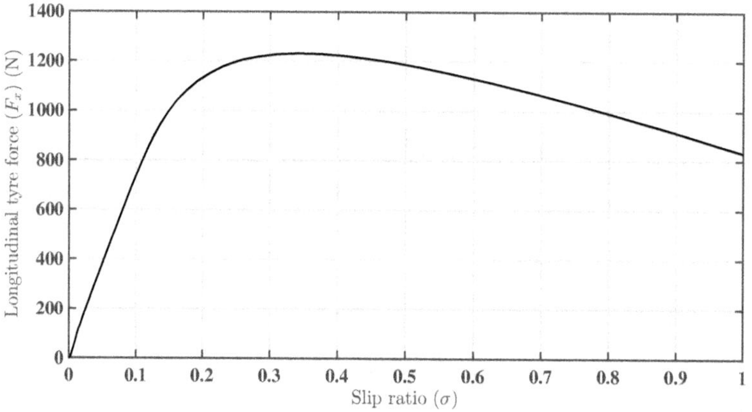

(b) Longitudinal tyre characteristics

Figure 3.5: Tyre characteristics of the NTTW vehicle

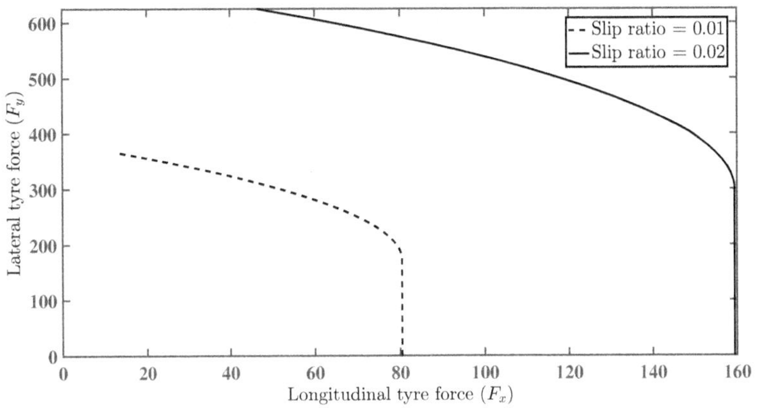

(a) Constant slip ratio

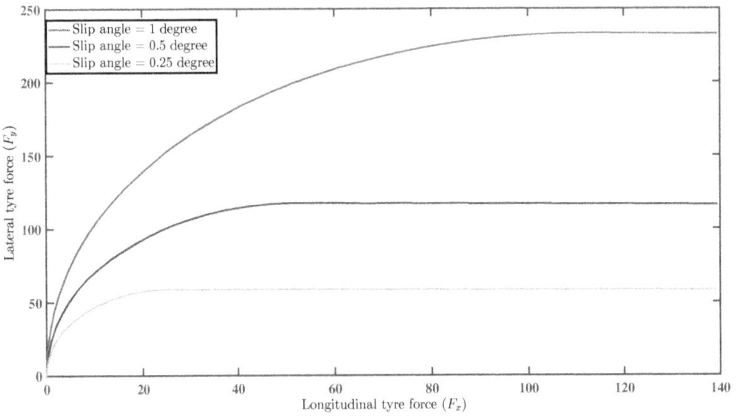

(b) Constant slip angle

Figure 3.6: Responses of the tyre model

3.5 Vertical dynamics model

Vertical load on tyres contribute to the evaluation of tyre slip and camber angles. Hence dynamic load shifting of tyre across longitudinal and transverse axis is crucial to be considered in the mathematical model of a vehicle dynamics. For example, when the load on the front wheel increases, a greater side force is generated for a given side slip and camber angles. The author in [6] has considered load transfer effect across the rear module during turning maneuver through the multi-body lateral dynamics model. This is particularly important during the transient phase of a maneuver as there is a peak in lateral acceleration. However, the effects of load transfer about front and rear axles during acceleration and braking motion are omitted. In most of the literature [31], [47], [45], the vertical load on the front tyres are assumed to be constant.

In this study, the variation of dynamic loading on tyres due to longitudinal, lateral and tilting motions are derived and included in dynamics equations. Figure 3.2 and Figure 3.3 depict the forces considered to calculate load transfer during various maneuvers. The dynamic vertical force (F_{z_i}) on each tyre is derived as shown in equations (3.13), (3.14) and (3.15). The suspension dynamics and coupling between pitch and roll of the vehicle are not considered.

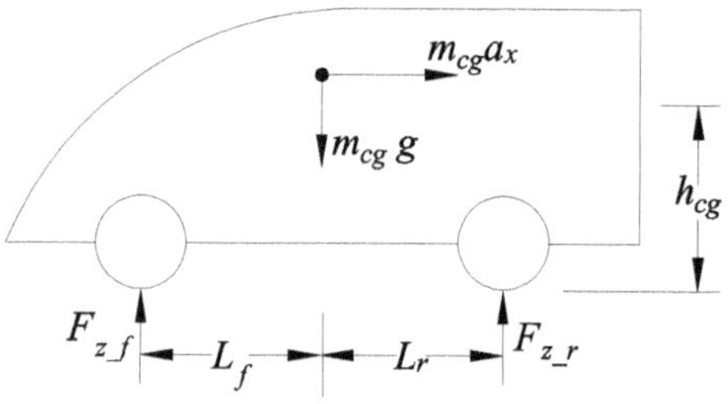

Figure 3.7: Load shift during acceleration

The scenario of load shifting between the front and rear axles during acceleration/braking is shown in Figure 3.7. Constructing the torque balance at the rear axle contact point, making the total moment ($\sum M_y$) around Y axis as zero.

$$\begin{aligned}\sum M_y &= 0\\ \Longrightarrow (L_f + L_r)F_{z\_f} - m_{cg}gL_r + m_{cg}a_x h_{cg} &= 0\\ \Longrightarrow F_{z\_f} &= \frac{L_r}{L}m_{cg}g - \frac{h_{cg}}{L}m_{cg}a_x\end{aligned} \tag{3.9}$$

Subtracting the front axle load from the total vehicle weight, the rear axle load is calculated as follows:

$$\begin{aligned}\sum F_z &= 0\\ \Longrightarrow (F_{z\_f} + F_{z\_r}) - m_{cg}g &= 0\\ \Longrightarrow F_{z\_r} &= \frac{L_f}{L}m_{cg}g + \frac{h_{cg}}{L}m_{cg}a_x\end{aligned} \tag{3.10}$$

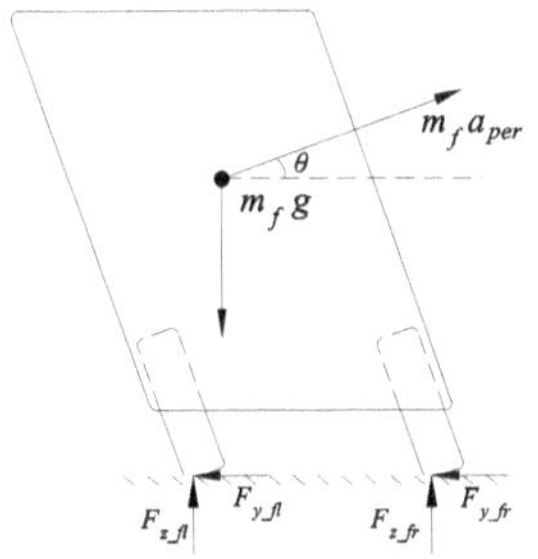

Figure 3.8: Load shift during cornering

The two axles are considered to be decoupled from each other, and a half-car model is used to calculate the lateral weight shift on each axle as shown in Figure 3.8. It shows the load shift on the front axle between left and right tyres due to cornering. During cornering, the lateral acceleration causes a weight shift to one side of each axle, whose distribution between the front and rear axles depends on the axle loads. The lumped mass of the front axle (m_f) is considered and the perceived acceleration (a_{per}) acts in the lateral direction at an angle of inclination (θ). The mass of the front axle m_f is calculated by equation (3.11).

$$m_f = \frac{F_{z\_fl} + F_{z\_fr}}{(g - a_{per}\sin\theta)} \tag{3.11}$$

Considering the torque balance equation at the point of contact between the ground and the front-left tyre, the lateral weight shift at the front-right tire is calculated by making total moment about X-axis ($\sum M_x$) as zero.

$$\sum M_x = 0$$
$$\implies F_{z\_fr} L_{tw} - m_f a_{per} \cos\theta h_{cg} - m_f (g - a_{per} \sin\theta) \frac{L_{tw}}{2} = 0$$
$$\implies F_{z\_fr} = m_f [\frac{(g - a_{per} \sin\theta)}{2} + \frac{a_{per} \cos\theta h_{cg}}{L_{tw}}]$$
$$\implies F_{z\_fr} = \frac{F_{z\_f}}{(g - a_{per} \sin\theta)} [\frac{(g - a_{per} \sin\theta)}{2} + \frac{a_{per} \cos\theta h_{cg}}{L_{tw}}]$$
$$\implies F_{z\_fr} = F_{z\_f} [\frac{1}{2} + \frac{a_{per} \cos\theta h_{cg}}{L_{tw}(g - a_{per} \sin\theta)}] \tag{3.12}$$

Substituting the total front axle mass from equation (3.11) and front axle load from equation (3.10) into equation (3.12), the front-right dynamic wheel load can be derived as follows:

$$F_{z\_fr} = m_{cg} [\frac{L_r}{L} g - \frac{h_{cg}}{L} a_x] [\frac{1}{2} + \frac{a_{per} \cos\theta h_{cg}}{L_{tw}(g - a_{per} \sin\theta)}] + \frac{M_\theta}{L_{tw}} \tag{3.13}$$

By analogy, the dynamic loads of the other two wheels can be computed as follows:

$$F_{z\_fl} = m_{cg} [\frac{L_r}{L} g - \frac{h_{cg}}{L} a_x] [\frac{1}{2} - \frac{a_{per} \cos\theta h_{cg}}{L_{tw}(g - a_{per} \sin\theta)}] - \frac{M_\theta}{L_{tw}} \tag{3.14}$$

$$F_{z\_r} = m_{cg} [\frac{L_f}{L} g + \frac{h_{cg}}{L} a_x] \tag{3.15}$$

Here, M_θ indicates the moment generated due to tilting of the vehicle body. The longitudinal acceleration (a_x) is represented by equation (3.16).

$$a_x = \dot{V_x} - V_y \dot{\psi} \tag{3.16}$$

3.6 NTTW vehicle body dynamics model

The driver of the motorcycle is solely responsible for maintaining the stability of the vehicle through steering input and shift in body weight. There is a need to relieve the driver of the task of maintaining stability by automatically making steering input and/or leaning the body of the vehicle in order to balance the lateral stability of the vehicle at the desired tilt angle for a forward speed of the vehicle. In order to overcome the tendency to overturn, NTTW vehicle is equipped with a tilting mechanism to learn the vehicle while taking a turn in much the similar manner as a motorcycle.

3.6.1 Lateral-tilt dynamics

The lateral direction is the sideways moving direction of the vehicle. There are two ways of looking at the lateral direction, with respect to the vehicle (local axis) and with respect to a fixed reference point (global axis). The former is often used when dealing with acceleration and velocity of the vehicle. The latter is used when the location information of the vehicle with respect to a starting or an ending point is desired. Researchers often find this direction more interesting than the longitudinal one since extreme values of lateral acceleration or lateral velocity can decrease vehicle stability and controllability [62].

Lateral dynamics equations are derived by equations (3.17),(3.18),(3.24) and (3.25). The states are the lateral velocity (V_y), the tilt angle (θ), the rate of change of tilt angle ($\dot{\theta}$) and the yaw rate ($\dot{\psi}$). The F_{x_fl}, F_{x_fr}, and F_{x_r} denote the front left, front right and rear longitudinal tyre forces. Whereas, F_{y_fl}, F_{y_fr}, and F_{y_r} represent the front left, front right and rear lateral tyre forces respectively. Both the front wheels are assumed to be steered with the same angle (δ_f) due to the narrow width of the vehicle. A tilting moment (T_{tilt}) is applied at CG of the NTTW vehicle to lean it during cornering maneuvers.

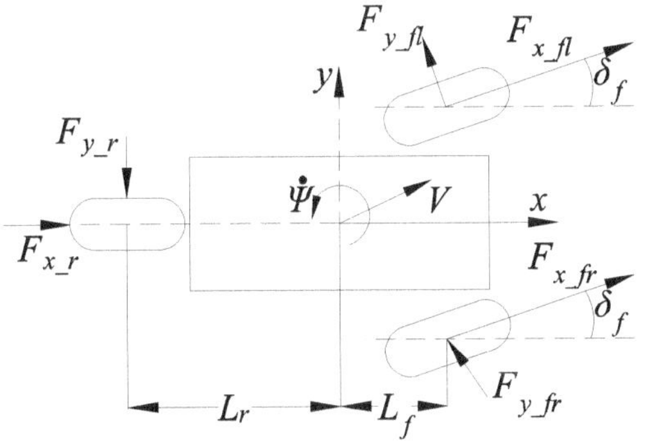

Figure 3.9: NTTW vehicle dynamic forces (Top view)

The Figure 3.9 showcases the lateral and longitudinal forces experienced at each tyre. Based on first principle method and balancing the force equations, lateral dynamics equations are derived as:

$$m_{cg}\dot{V}_y = F_{x\_fl}\sin\delta_f + F_{x\_fr}\sin\delta_f + F_{y\_r} + F_{y\_fl}\cos\delta_f + F_{y\_fr}\cos\delta_f - m_{cg}\dot{\psi}V_x + f(\theta) \quad (3.17)$$

$$f(\theta) = m_{cg}h_{cg}\ddot{\theta}\cos\theta - m_{cg}h_{cg}\dot{\theta}^2\sin\theta \tag{3.18}$$

$$\begin{aligned} I_{z\_veh}\ddot{\psi} = \quad & (F_{x\_fl}\sin\delta_f + F_{x\_fr}\sin\delta_f)L_f + (F_{y\_fl}\cos\delta_f + F_{y\_fr}\cos\delta_f)L_f - F_{y\_r}L_r + \\ & \tfrac{L_{tw}}{2}(F_{x\_fr}\cos\delta_f - F_{x\_fl}\cos\delta_f) + \tfrac{L_{tw}}{2}(F_{y\_fl}\sin\delta_f - F_{y\_fr}\sin\delta_f) \end{aligned} \tag{3.19}$$

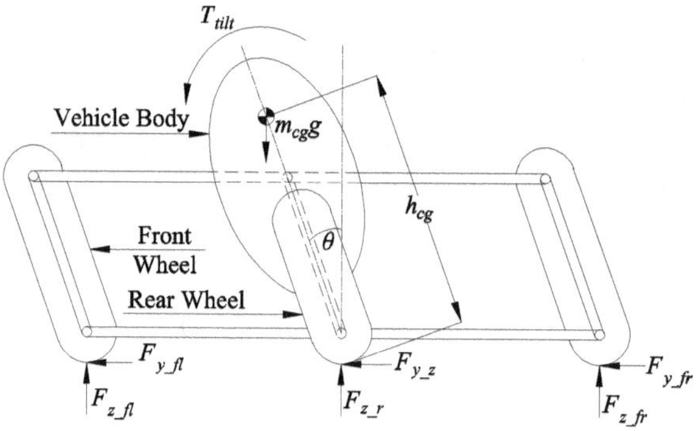

Figure 3.10: NTTW vehicle dynamic forces (rear view)

The tilting body of the NTTW vehicle and the forces that each of the tyres experience is shown in Figure 3.10. Following Newtonian's rule, the tilting moment generated by the NTTW vehicle ($I_{x\_veh} \times \ddot{\theta}$) must balance the total moments generated by the various tyre forces (vertical and lateral) and the input torque, while maneuvering. The final equation is derived as:

$$I_{x\_veh}\ddot{\theta} = F_{z\_fr}[\frac{L_{tw}}{2} + h_{cg}\sin\theta] - F_{z\_fl}[\frac{L_{tw}}{2} - h_{cg}\sin\theta] + F_{z\_r}h_{cg}\sin\theta - F_y h_{cg}\cos\theta + T_{tilt} \tag{3.20}$$

3.6.2 Longitudinal dynamics

The longitudinal dynamics concerns itself with the forward moving direction of the vehicle. There are two different ways of looking at the forward direction, one with respect to the vehicle body itself (local axis), and another with respect to a fixed reference point (global axis).

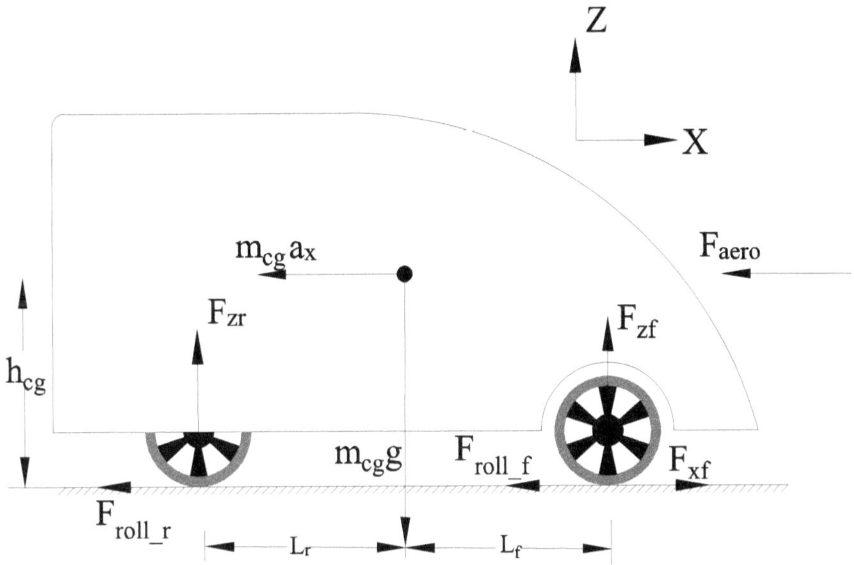

Figure 3.11: Schematic for longitudinal vehicle dynamics for NTTW vehicle

Equations (3.21),(3.22) and (3.23) derive the longitudinal dynamics. The states are the longitudinal velocity (V_x), the left wheel angular velocity (ω_fl) and the right wheel angular velocity (ω_fr) as only these two wheels are driven.

$$m_{cg}\dot{V_x} = F_{x\_fl}\cos\delta_f + F_{x\_fr}\cos\delta_f + F_{x\_r} - F_{y\_fl}\sin\delta_f - F_{y\_fr}\sin\delta_f + m_{cg}\dot{\psi}V_y - F_{aero} - F_{grad} \quad (3.21)$$

In this work, the driving force is provided by two front in-wheel hub motors, whereas the rear wheel is assumed to be purely rolling. The wheel dynamics are derived by balancing the moment equations using the driver's torque and the tyre resisting forces. F_{roll_i} denotes the rolling resistance of the tyres. T_{d_fl} and T_{d_fr} are the torques assumed to be provided by the in-wheel hub motors. The rolling resistance of front left and right wheels are represented by F_{roll_fl} and F_{roll_fr} respectively.

$$I_{\omega\_fl}\dot{\omega}_{fl} = T_{d\_fl} - T_{b\_fl} - rF_{x\_fl} - rF_{roll\_fl} \quad (3.22)$$

$$I_{\omega\_fr}\dot{\omega}_{fr} = T_{d\_fr} - T_{b\_fr} - rF_{x\_fr} - rF_{roll\_fr} \quad (3.23)$$

The external forces acting on a vehicle are aerodynamic resistance (F_{aero}), the rolling resistance of the tyres (F_{roll}) and the gradient resistance (F_{grad}). The calculation of the vehicle resistance forces is given in Appendix A.

3.7 Trajectory of the vehicle

To evaluate the directional stability of the vehicle, the trajectory of the vehicle's movement on the global axis can be monitored. The vehicle trajectory with respect to global X and Y coordinates are derived from equations (3.24) and (3.25) to investigate about the instantaneous position of the vehicle [62].

$$X = \int [V_x \cos\psi(t) - V_y \sin\psi(t)]dt \tag{3.24}$$

$$Y = \int [V_x \sin\psi(t) + V_y \cos\psi(t)]dt \tag{3.25}$$

3.8 Actuator modeling

It is assumed that the driving force is provided by two front in-wheel hub electric motors. The front wheels are assumed to be steered by a steer-by-wire mechanism. Another actuator is assumed to be used to lean the vehicle body about its longitudinal axis. All the four actuators, i.e. the two in-wheel hub motors, the steer-by-wire actuator and the tilt actuator, are modeled as first order transfer functions. There are four actuators in total, the time constants of these actuators used in this work are listed in Table 3.2. The tilt actuator and the in-wheel hub motors are assumed to be faster in response than the steering actuator. Moreover, it is critical that the limits of operation of various actuation systems are maintained while complex maneuvers are commanded. Hence, the actuator inputs, viz. δ_f, T_{tilt} and T_{d_i} are saturated at ± 30 deg, ± 150 Nm and ± 500 Nm respectively. The inputs are bounded by these values for a realistic analysis of the performance of the controller. Further, the maximum operating speed of the NTTW vehicle is considered to be 75 kmph.

Table 3.2: Time constants of the actuators

Actuator	Time constant
Steering Actuator	20 $msec$
Tilt Actuator	2 $msec$
Two in-wheel hub motors	2 $msec$

3.9 Parametrization of NTTW vehicle

The design of NTTW vehicle chassis has been documented in Appendix A. A realistic design specification for NTTW vehicle has been chalked out after a thorough study made on past research carried out in academia and industries on NTTW vehicles. Table 3.3 lists the pertinent parameters with values which are utilized in the mathematical model for simulation work.

Table 3.3: Parameter definition

Parameter	Symbol	Value	Unit
Mass	m_{cg}	450	kg
Wheelbase	L	1.7	m
Distance of center of gravity to front wheels	L_f	0.6	m
Distance of center of gravity to rear wheels	L_r	1.1	m
Trackwidth	L_{tw}	0.85	m
Height of center of gravity	h_{cg}	0.6	m
Diameter of tyre	d_{tyre}	0.55	m
Tilt moment of inertia of vehicle	I_{x_veh}	111.5	kgm^2
Yaw moment of inertia of vehicle	I_{z_veh}	243	kgm^2
Cornering stiffness of front left wheel	C_{α_fl}	13550	$Nrad^{-1}$
Cornering stiffness of front right wheel	C_{α_fr}	13550	$Nrad^{-1}$
Cornering stiffness of rear wheel	C_{α_r}	14592	$Nrad^{-1}$
Camber stiffness of front left wheel	C_{λ_fl}	2705	$Nrad^{-1}$
Camber stiffness of front right wheel	C_{λ_fr}	2705	$Nrad^{-1}$
Camber stiffness of rear wheel	C_{λ_r}	2918	$Nrad^{-1}$
Longitudinal stiffness of front left wheel	C_{σ_fl}	8130	$Nrad^{-1}$
Longitudinal stiffness of front right wheel	C_{σ_fr}	8130	$Nrad^{-1}$
Longitudinal stiffness of rear wheel	C_{σ_r}	8755	$Nrad^{-1}$

3.10 Open loop test cases

Open loop response of a system enables to analyze its characteristics to establish its basic understanding. Hence initially, simulations are performed to analyze the dynamics of NTTW vehicle model in open loop. The behavior of individual sub-systems, i.e., lateral-tilt and longitudinal dynamics are analyzed by monitoring their responses when a step input is introduced into the system.

A step change of 0.1 degrees of δ_f and 30 Nm of T_{tilt} are provided as inputs to the lateral-tilt dynamics sub-system at 5 sec of simulation time (Figure 3.12). It can be observed from the evolution of the lateral-tilt states shown in Figure 3.13, that as time progresses, the states move towards instability. This behavior is due to the fact that the eigenvalues of the lateral-tilt dynamics have unstable poles and the analysis of eigenvalues of the NTTW vehicle dynamics is presented in 4.6.

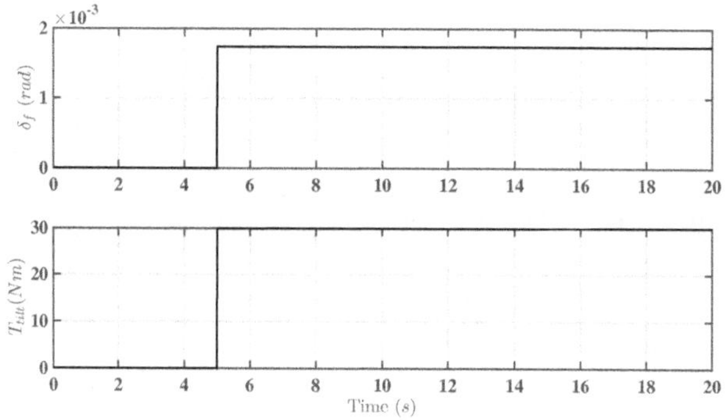

Figure 3.12: Open-loop step inputs to lateral-tilt dynamics of NTTW vehicle

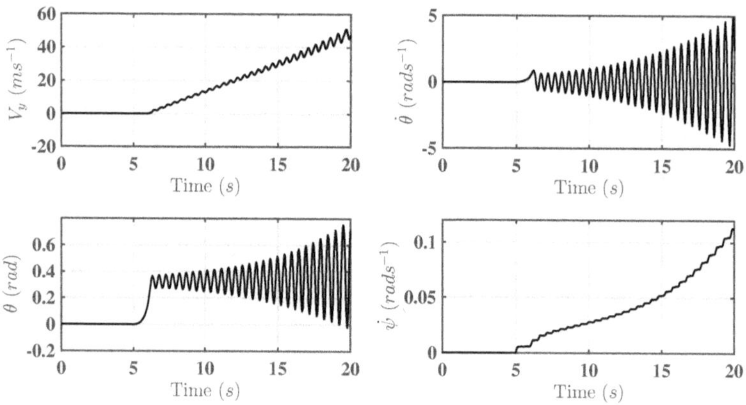

Figure 3.13: Lateral-tilt states in open loop

Similarly, open loop test is performed on longitudinal dynamics by providing a step change of 150 Nm to 20 Nm to the drive torques (T_{d_fl} and T_{d_fr}) as shown in Figure 3.14. It can be observed from the plot of V_x in Figure 3.15 that due to lack of control the speed keeps on increasing and hence tends towards instability.

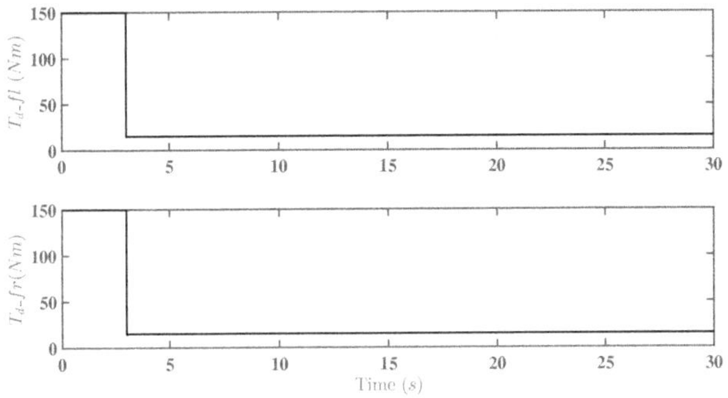

Figure 3.14: Open-loop step inputs to longitudinal dynamics of NTTW vehicle

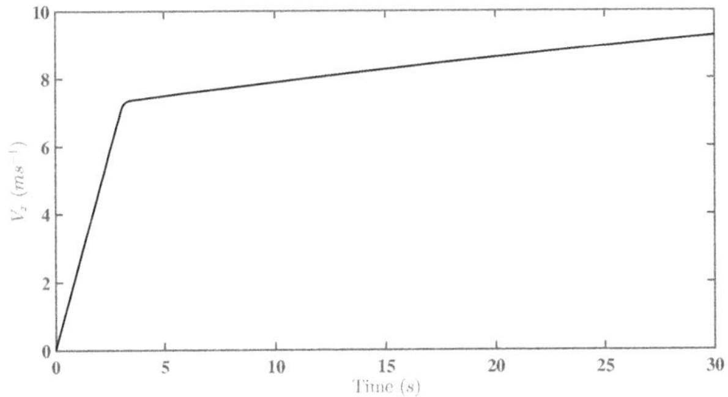

Figure 3.15: Longitudinal state in open loop

Thus, the open loop tests provide a motivation for appropriate control design of the NTTW vehicle to achieve the desired handling and stability performance.

3.11 Summary

i. A detailed seven DoF non-linear NTTW vehicle model is formulated. The lateral-tilt dynamics consists of 4 DoF (V_y, $\dot{\theta}$, θ and $\dot{\psi}$) and the longitudinal dynamics comprises of three DoF (V_x, ω_{fl} and ω_{fr}). The lateral-tilt and longitudinal dynamics are interlinked through the non-linear tyre model (slip ratio, slip angle and tilt angle) and the vertical dynamics model (vertical tyre forces). It is observed that there is a strong coupling within the dynamics which adds to the complexities of the complete platform.

ii. The NTTW vehicle is considered to have two driven front wheels and one non-driven rear wheel. The driven wheels house in-wheel hub electric motors which provide the driving force. The front wheels are assumed to be steered by a steer-by-wire mechanism. Another actuator is assumed to be used to lean the vehicle body about its longitudinal axis. All the four actuators, i.e. the two in-wheel hub motors, the steer-by-wire actuator and the tilt actuator, are modeled as first order transfer functions.

iii. The mathematical model is intended for understanding the behavior of the NTTW vehicle during dynamic maneuvering and then to further assist in the design of active stability control systems and is used in this research work.

Chapter 4

NTTW Vehicle Integrated dynamics Control Design: Problem Formulation

4.1 Introduction

The goal of this study is to address vehicle dynamics stability issues typical to the proposed NTTW platform. It is aimed at analysis of vehicle dynamics along longitudinal (traction control), lateral (steering maneuvers, yaw control with optimum tilting) and vertical (dynamic loadings) directions individually. Also, the inter-relationship among the aforementioned functions is analyzed and an optimized integrated control strategy is designed for active vehicle dynamics under various dynamic situations. The integration effects are manifested due to strongly coupled vehicle motions through tyre dynamics. A nonlinear seven degree of freedom mathematical model of NTTW vehicle platform is formulated in Chapter 3 which is the plant for the controller. This chapter describes the control system design procedure applied in this thesis work. It first provides the definition of certain parameters, i.e. static envelope and perceived acceleration which are required in analyzing the handling behavior of NTTW platform. A set of control objectives are drawn based on the functionality of the NTTW platform and typical vehicular stability conditions. Then, a set of test case of vehicle maneuvers is reported which are utilized to analyze the proposed controllers in Chapters 5 through 10.

4.2 Integrated vehicle control strategy

The interdependencies amongst longitudinal, vertical, lateral and tilt dynamics are explained in Figure 2.10. The longitudinal, lateral and yaw motions of a vehicle are primarily determined by tyre friction forces at the contact patch between tyre and road. The optimum driving dynamics can only be achieved when tyre forces on all the wheels and in all directions can be influenced and controlled precisely. The equilibrium as required for the stability of NTTW vehicle is not fixed as velocity varies with time and the desired tilt angle changes as velocity is changing. The leaning of the vehicle body at a desired angle (which is continually changing) demands incorporation of an active control system to ensure the tilt stability without affecting the handling behavior of the vehicle. It is, thus, difficult to deal with integrated dynamics control system for stability analysis of narrow tilting three wheeled vehicle due to the interactions among subsystems. This results in complexity in the design of controller as each subsystem has its own envelope of operating requirements. Also, large space of multiple pairings and operating conditions while coordinating functionalities among the subsystems further complicates the control design.

An overall concept of the integrated vehicle control strategy which is applied in this work is provided in Figure 4.1. There are two separate controllers, one for lateral-tilt dynamics and the other for longitudinal dynamics. The desired yaw rate and tilt angle are monitored by lateral-tilt dynamics controller, whereas desired longitudinal speed and angular velocities of each driven wheel are monitored by longitudinal dynamics controller. The prime objective is to achieve the desired trajectory of the driver while maintaining vehicle stability and comfort of the occupants. The vehicle is assumed to be equipped with a tilting actuator and a steer-by-wire mechanism to follow the input commands from the lateral-tilt controller. Similarly, the longitudinal controller input commands are followed by the two in-wheel hub motors, as described in Section 3.8.

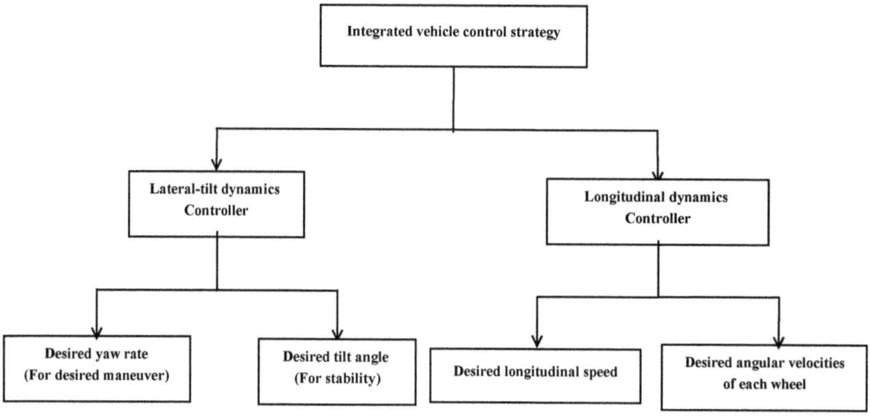

Figure 4.1: The concept of Integrated Vehicle Stability Controller Design

Table 4.1: Static envelope of model

Rc	Longitudinal speed (m/s)							
	Vx (m/s)	1	2	3	7	12	18	22
3.5	ay (m/s^2)	0.29	1.14	2.57	14	41.14	92.57	138.29
	δSW (deg)	27.91	28.09	28.4	30.89	36.81	48.01	57.97
	θ (deg)	1.69	6.66	15.02	81.81	240.4	540.93	808.1
14	ay (m/s^2)	0.07	0.29	0.64	3.5	10.29	23.14	34.57
	δSW (deg)	6.98	7.02	7.1	7.72	9.2	12	14.49
	θ (deg)	0.41	1.69	3.74	20.45	60.13	135.22	202.01
45	ay (m/s^2)	0.02	0.09	0.2	1.09	3.2	7.2	10.76
	δSW (deg)	2.17	2.19	2.21	2.4	2.86	3.73	4.51
	θ (deg)	0.12	0.53	1.17	6.37	18.7	42.07	62.88

4.3 Static envelope for the controller

The three wheeled vehicles are more susceptible to rollover stability than the four wheeled vehicle. As the lateral acceleration of vehicle increases, the tyre vertical force decreases until it becomes zero. The rollover is initiated when inner wheel lifts off the ground. The rollover stability condition for three wheeled vehicle with two front wheels is discussed in [5]. The threshold acceleration value is represented by equation (4.1).

$$a_{y\_ro} = \frac{L_{tw}}{2h_{c.g}} \frac{L_r g}{L} \tag{4.1}$$

The threshold rollover acceleration value (a_{y_ro}) of a three wheeled vehicle is dependent on geometric parameters of the vehicle such as, the vehicle track width, height of the location of CG of the vehicle and longitudinal position of CG of the vehicle. Based on the geometric parameters considered in this work, the value of a_{y_ro} is calculated as $4ms^{-2}$. This value falls under medium lateral acceleration range as given in [25]. Hence, the driver inputs are limited to these values for a better analysis of the performance of the controllers. For example, longitudinal speed of around 25 kmph (7 ms^{-1}) can be achieved for a minimum turning radius of 14 m. In this case, the maximum steering angle that could be exerted by the driver is 7.7 degree. Some of these values that constitute the static envelope of the model is shown in Table 4.1. The shaded region of this table represents all areas where the lateral acceleration exceeds $4ms^{-2}$. Though, under dynamic conditions, the envelope may be exceeded in transients.

4.4 Reference Model

A reference model converts the drivers intent (steering wheel angle and reference speed) into desired signals for the controllers (tilt angle, yaw rate, speed and angular velocities of tyres). As shown in Figure 4.1, there are two separate control paths, one for lateral-tilt and another for longitudinal dynamics. Hence, the desired signals are generated for these two separate paths to ultimately achieve an integrated desired path. The integrated (lateral-tilt and longitudinal) desired path is compared with the actual path of the non-linear model to generate necessary corrective action to maintain the desired trajectory.

4.4.1 Lateral-tilt reference parameters

The reference model for the lateral-tilt dynamics controller converts driver's command (steering wheel angle, δ_{sw}) into desired signals ($\dot{\psi}_{des}$ and θ_{des}). Driver's required radius of curvature (R_{des}) is calculated using the driver's steering angle (δ_{sw}). Consequently, the desired yaw rate ($\dot{\psi}_{des}$) is calculated using R_{des} and the vehicle longitudinal speed. This is mathematically represented as [31]:

$$\begin{aligned} R_{des} &= \frac{L}{\delta_{sw}} \\ \Rightarrow \dot{\psi}_{des} &= \frac{V_x}{R_{des}} \end{aligned} \tag{4.2}$$

Where V_x is the current longitudinal speed obtained from the longitudinal dynamics model.

Desired tilt angle (θ_{des})

The desired tilt angle (θ_{des}) is the angle at which the centrifugal force and gravitational force balance out each other for a specific turn of radius R_{des}. This results in zero perceived lateral acceleration. The vehicle begins yawing before it starts tilting and yaw can be utilized to determine the desired tilt angle which is given by equation (4.3). The desired tilt angle is derived from the current vehicle yaw rate and speed [39] as:

$$\theta_{des} = \arctan(\frac{\dot{\psi} V_x}{g}) \tag{4.3}$$

Where the $\dot{\psi}$ and V_x signals are current values of the yaw rate and longitudinal speed are obtained online from the lateral-tilt and longitudinal dynamics model respectively. The inclusion of instantaneous yaw rate ($\dot{\psi}$) to calculate the θ_{des} aids in minimizing the initial peak of tilting torque [44].

Perceived lateral acceleration (a_{per})

The perceived acceleration (a_{per}) is estimated as the lateral acceleration along the Y' axis in conjunction with the influence of gravity and the rotating movement of the vehicle body [51]. The perceived lateral acceleration (a_{per}) is resultant acceleration at CG of the vehicle along the Y axis (Figure 3.2) and it tends to zero ensuring lateral stability of the vehicle [51]. Also, it signifies that the ride is comfortable for the occupants. The relationship to find a_{per} is shown in equation (4.4) [39].

$$a_{per} = a_y \cos\theta + h_{cg}\ddot{\theta} - g\sin\theta \tag{4.4}$$

Where,

$$a_y = \dot{V}_y + V_x\dot{\psi}$$

In this work, the a_{per} is one of the important parameters which is monitored to analyze the performance of the proposed control strategies to ascertain lateral stability and comfort level of the riders.

4.4.2 Longitudinal reference parameters

For the longitudinal dynamics controller, driver's speed command is considered as the desired longitudinal speed (V_{x_des}). In addition, in case of a multivariable controller, all the states (V_x, ω_{fl} and ω_{fr}) of the longitudinal dynamics are fed back to the controller to achieve desired characteristics. The desired angular velocities of the front left and front right wheels (ω_{fl_des} and ω_{fr_des}) are also considered and calculated using the relation given by equation (4.5).

$$\omega_{i\_des} = \frac{V_x}{\frac{d_{tyre}}{2}} \tag{4.5}$$

where, 'i' denotes the front left (fl) or the front right (fr) wheel.

4.5 Control system objectives

Since the main aim is to achieve the integrated chassis stability, a set of control objectives are drawn based on the functionality of the NTTW platform and typical vehicular stability conditions. These control system objectives are hence similar to that of required in a practical scenario when the controller is deployed on a real vehicle. The control objectives are:

i. To track the desired maneuver as commanded by the driver in terms of desired yaw rate ($\dot{\psi}_{des}$), desired tilt (θ_{des}) and desired longitudinal speed (V_{x_des})

ii. To monitor the a_{per} by keeping the transient to be a_{per} minimum and steady state value to be maintained at zero

iii. Minimization of required transient and steady-state tilt torque

iv. To ensure the tyres are functioning within their traction capabilities in terms of its loading during a maneuver

v. To maintain robustness of the controller in presence of parametric changes

vi. To be less sensitive to external disturbances

vii. To design simple implementable controller matching available vehicular embedded systems

4.6 Test maneuvers for evaluating vehicle handling and performance

From a control system perspective, the NTTW vehicle (plant) and controller form a loop where the controller is responsible for the stability of the NTTW vehicle (plant). In such a control loop, some disturbances act on the vehicle (such as road resistance forces) and the others act on the controller (such as matched input disturbance). In terms of lateral-tilt dynamics, the actuating variables (i.e. steering wheel angle and tilting torque) are corrected by the controller. Similarly, for longitudinal dynamics, the actuating variables, i.e. front wheel drive torques are corrected by the controller. Thus, the controller corrects the difference between the desired and actual paths and speeds for lateral-tilt and longitudinal dynamics respectively. A vehicle is considered to have good handling characteristics when [25]:

i. A good correlation exists between the steering wheel and the lane-change behavior of the vehicle

ii. Little or no change in the course of the vehicle against the external disturbances acting on the vehicle is exhibited

iii. A reasonable information of feedbacks on actuating variables, to predict the behaviors of the vehicle, is available for reacting correctly

iv. A high lateral acceleration limit of the vehicle is available

Many test maneuvers are used in the industry which provides insight into the dynamic behavior of the vehicle under different driving conditions. A detailed study has been documented in [25]. As per the scope of the research work, a comprehensive evaluation of the dynamic characteristics of

NTTW vehicle can be carried out by examining the results obtained from the selected maneuvers in order to establish the effectiveness of different chassis control systems under study. The NTTW vehicle is required to be evaluated with respect to three main performance characteristics, i.e., handling, stability and path-following. Four test maneuvers i.e. constant radius turn, double lane change, accelerating while cornering and decelerating while cornering are used in this work. The criteria for selection of such test maneuvers are explained in subsequent sub-sections. The chosen test maneuvers are described in the Tables 4.2 through 4.3. The aim is to have a single set of gain matrix for lateral-tilt and longitudinal control to achieve required performance across a wider envelope of operation. The trajectories mentioned in Table 4.2 are quite difficult and diverse for investigations than the ones dealt in the literature [41], [51]. For example, 500 m of turn radius is considered in [44] which is quite high compared to the ones used in this thesis work.

Table 4.2: Maneuvers for proposed integrated multivariable controllers

Test Maneuvers	Radius of turn (m)	Velocity (kmph)
Double lane change	10	10
Constant radius turn	35	30
	75	30
	200	30
	75	40
Accelerating while cornering	170	20 to 75
Deceleration while cornering	135	30 to 25

Table 4.3: Maneuvers for Comparison between controllers

Test Maneuvers	Radius of turn (m)	Velocity (kmph)
Double lane change	23	20
Double lane change with external disturbances	23	20

As shown in Table 4.4, matched input disturbances are added to the control inputs (δ_f and T_{tilt}) of the lateral-tilt dynamics. The F_{aero} and F_{grad} are major external resistance forces acting on the longitudinal dynamics of the NTTW vehicle. Thus, addition of these tests provides a realistic representation of the external disturbances experienced by the NTTW vehicle.

Table 4.4: Various disturbances to test performance of the controllers

Sr.	Disturbances	Nature	Time
1.	20% of maximum inputs (δ_f and T_{tilt}) at the input channel	Constant	From 14 s to 30 s
3.	5% of maximum F_x added at F_{aero}	Sinusoidal	From 14 s to 30 s
4.	5 degree gradient force at F_{grad}	Constant	From 12 s to 18 s

The subsequent sub-sections provide a detailed description of the vehicle maneuvers chosen to analyze the performance of the proposed controllers.

4.6.1 Constant radius of turn

The steady state handling performance is concerned with the directional behavior of a vehicle during a turn under non time-varying conditions [65], e.g. a steady state turn is a vehicle negotiating a curve with a constant radius of turn at a constant forward speed. It determines how promptly the vehicle reacts to a steering input and how quickly it can settle to a new equilibrium. Constant radius turn tests are simulated to analyze the performance of the stability controllers for the NTTW vehicle and are explained in relevant sections in Chapters 5 through 10.

4.6.2 Double lane change

The performance measure to analyze a vehicle's transient dynamics with a constant forward speed is done through pulse impulse test such as double-lane change maneuver. Initially, the vehicle moves in a straight line and then, the driver transients to take a turn of given radius of curvature. After a while, the driver changes the direction of travel to a straight line until the end of simulation. All this while, the longitudinal speed is maintained at constant reference.

4.6.3 Acceleration and deceleration while cornering

A three wheeled vehicle is less stable as compared to a four wheeled vehicle. [5] analyzed rollover stability while accelerating/decelerating for three wheeled vehicle, which has a larger overturning moment and consequently, lower rollover speed limits. Hence, these two tests are very important for evaluating the stability performance of NTTW vehicle. For investigation of controller behavior during acceleration/deceleration while cornering, it is important to differentiate between large and small magnitudes. In this work, low deceleration rate (0.05 ms^2) and moderate acceleration rate (1 ms^2) are considered during simulations. The driver (speed command) accelerates from 0 to 20 kmph ($\approx 5.5ms^{-1}$) to 75 kmph ($\approx 21ms^{-1}$) while taking a step-steer turn of a fixed turning radius of curvature. The braking system is not included in the scope of the thesis, hence a case of minimal deceleration is considered for analysis. The driver (speed command) accelerates from 0 to 25 kmph ($\approx 7ms^{-1}$) and decelerates to 22 kmph ($\approx 6ms^{-1}$) while taking a step-steer turn of fixed turning radius of curvature.

4.6.4 Plots

The plots are graphical representations of various vehicle responses during stability investigations. The list of useful plots of different parameters of NTTW vehicle to analyze its behavior during dynamic maneuvers are provided below.

i. Control inputs: The required control inputs for a particular maneuver are plotted. The control inputs for lateral-tilt dynamics are the steering angle (δ_f) and the tilting torque (T_{tilt}), whereas, for the longitudinal dynamics the inputs are the front left and right drive torques (T_{d_fl} and T_{d_fr}).

ii. Lateral-tilt states: Graphs for the evolution of the states of lateral-tilt dynamics are reported for each controller. The controlled states (θ and $\dot{\psi}$) are plotted with desired and actual states to test the tracking performance. The other uncontrolled states (V_y and $\dot{\theta}$) are also plotted to ensure that they are within bounds. A good correlation should exist between the pattern of V_y and the desired steering wheel of the vehicle.

iii. Longitudinal states: Graphs representing the evolution of the states of the longitudinal dynamics are also reported for each controller. The controlled states (V_x, ω_{fl} and ω_{fr}) are plotted with desired and actual states to test the tracking performance.

iv. Vehicle accelerations: Various vehicle body accelerations, viz. a_x, a_y and a_{per} are also plotted to monitor their values.

v. Slip angles, slip ratios: These parameters are crucial as they form the basis for the generation of lateral and longitudinal tyre forces and are plotted for examinations.

vi. Lateral, longitudinal and vertical tyre forces: All the three tyre forces (front left, front right and rear) are monitored.

vii. Global XY trajectory of NTTW vehicle: Actual and desired vehicle trajectories for different control strategies are reported to monitor how closely the desired trajectory is being followed.

4.7 System Linearization

The NTTW vehicle model is represented as a seven DoF non-linear mathematical equations as explained in detail in Chapter 3. These non-linear equations can be approximated by a linear form that is more beneficial for control design and practical implementation, especially when deploying the controller in embedded systems. The simplest approach to linearize the model dynamics is by performing a linear approximation of the non-linear system model around a desired operating point. This linear model will be valid in small regions around the operating point. Then, to control the system dynamics, traditional linear control methods will be applied. Such a technique is frequently applied in practical scenarios. The linearized model is used to design controllers that are described in Chapter 6 and Chapter 8.

The lateral-tilt dynamics and longitudinal dynamics are linearized independently by performing linear approximation around suitable operating points (about different longitudinal velocity values, such as $V_x = 30$ kmph and $\theta = 0$ deg) using linear analysis tool of Matlab. The linearized models are validated using simulation by giving same inputs to linear and non-linear models and then comparing the output responses. The validation method of the linearized models are explained in detail in subsequent sections.

4.7.1 Linearization of lateral-tilt dynamics

In this section, linearization of lateral dynamics block is explained in detail. System matrices of the lateral-tilt dynamics are obtained by linearization of the non-linear open loop mathematical model. The eigenvalues of lateral-tilt dynamics for $V_x = 20kmph$ are: [-49.6253;3.0733;-10.3931;-4.8836]. The linearized NTTW vehicle model (used to design the controller) is validated with the nonlinear model by providing the same input to both the models and observing their responses. It can be observed from the open-loop response given in Figure 3.13 and the eigen values, that the system is inherently unstable. Hence, a controller based on basic PID strategy is employed for this validation for both the models.

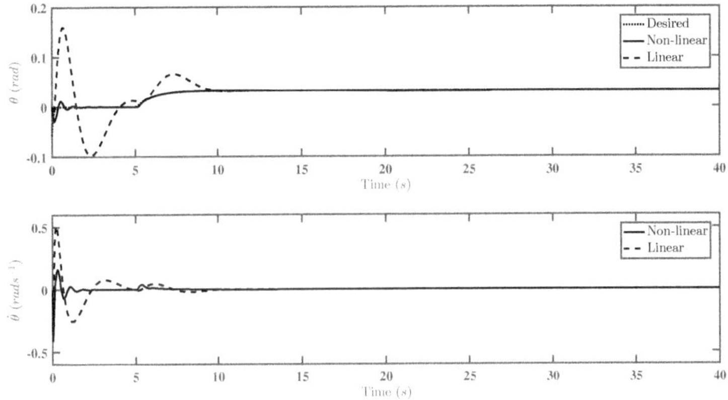

Figure 4.2: Comparison of tilt states between the non-linear and linear model

The tilt angle (θ) and rate of change of tilt angle ($\dot{\theta}$) are monitored as those are the states that are being controlled by the baseline controller.

As can be seen from Figure 4.2, the steady-state values match for the non-linear and linear models, though the transient responses are comparatively different. The non-linear NTTW vehicle model with lateral and tilt dynamics is represented in linear state space form as:

$$\dot{\mathbf{x}}_{lat} = A_{lat}\mathbf{x}_{lat} + B_{lat}\mathbf{u}_{lat} \tag{4.6}$$

$$\mathbf{y}_{lat} = C_{lat}\mathbf{x}_{lat} + D_{lat}\mathbf{u}_{lat}$$

where, $\mathbf{x}_{lat} = \begin{bmatrix} V_y & \dot{\theta} & \theta & \dot{\psi} \end{bmatrix}^T$ is state vector, $\mathbf{u}_{lat} = \begin{bmatrix} \delta_f & T_{tilt} \end{bmatrix}^T$ is control vector and $\mathbf{y}_{lat}$ denotes the measured outputs, $\mathbf{y}_{lat} = \begin{bmatrix} \dot{\theta} & \theta & \dot{\psi} \end{bmatrix}^T$. The unmeasured state, V_y is synthesized using a suitable estimator design presented in Chapter 7.

The non-linear lateral-tilt dynamics model is linearized at different operating points. Some of the operating points and the resultant linearized matrices are listed below.

Case 1: $V_x = 20kmph$

$$A_{lat} = \begin{bmatrix} -41.32 & 0 & 31.2 & -21.5 \\ 40.8 & 0 & -21.11 & 15.78 \\ 0 & 1 & 0 & 0 \\ -12.06 & 0 & 0 & -20.51 \end{bmatrix}, \quad B = \begin{bmatrix} 147.72 & -0.0054 \\ -145.83 & 0.0090 \\ 0 & 0 \\ 66.35 & 0 \end{bmatrix}$$

$$C = \begin{bmatrix} 1 & 0 & 0 & 0 \\ 0 & 0 & 0 & 1 \\ 0 & 1 & 0 & 0 \\ 0 & 0 & 1 & 0 \end{bmatrix}, \quad D = \begin{bmatrix} 0 & 0 \\ 0 & 0 \\ 0 & 0 \\ 0 & 0 \end{bmatrix}$$

Eigen values are: [-49.6253; 3.0733; -10.3931; -4.8836]

Case 2: $V_x = 30kmph$

$$A = \begin{bmatrix} -28.22 & 0 & 33.2 & -7.9 \\ 27.86 & 0 & -27.11 & -0.11 \\ 0 & 1 & 0 & 0 \\ 0.08 & 0 & 0 & -14.05 \end{bmatrix}, \quad B = \begin{bmatrix} 147.72 & -0.0054 \\ -145.83 & 0.0090 \\ 0 & 0 \\ 66.35 & 0 \end{bmatrix}$$

Eigen values are: [-27.0167; 2.6472; -3.7908; -14.1153]

Case 3: $V_x = 55kmph$

$$A = \begin{bmatrix} -22.58 & 0 & 31.2 & -9.9 \\ 22.3 & 0 & -21.11 & -0.86 \\ 0 & 1 & 0 & 0 \\ 0.07 & 0 & 0 & -11.24 \end{bmatrix}, \quad B = \begin{bmatrix} 147.72 & -0.0054 \\ -145.83 & 0.0090 \\ 0 & 0 \\ 66.35 & 0 \end{bmatrix}$$

Eigen values are: [-21.0089; 2.5507; -4.0221; -11.3400]

4.7.2 Linearization of longitudinal dynamics

Similar to the previous section, the linearization and validation of longitudinal dynamics are performed by feeding the same inputs to the linear and non-linear longitudinal closed loop dynamics.

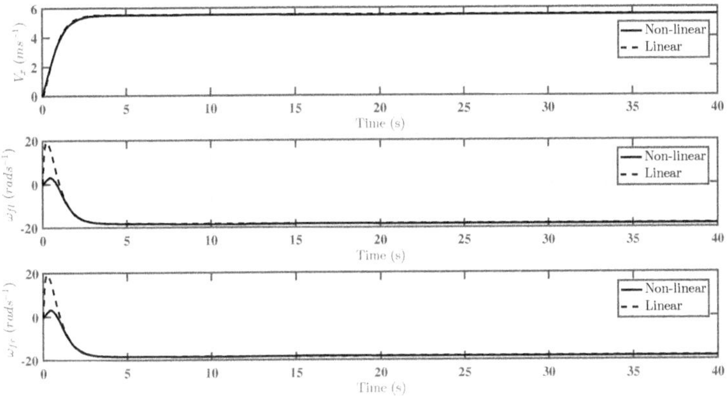

Figure 4.3: Comparison of longitudinal states between the non-linear and linear model

The longitudinal dynamics states, forward speed (V_x) and angular velocities (ω_{fl} and ω_{fr}) of the wheels are monitored here. As can be seen from Figure 4.3, the steady-state values match for the non-linear and linear models. The non-linear NTTW vehicle model with longitudinal dynamics is represented in linear state space form as:

$$\dot{\mathbf{x}}_{long} = A_{long}\mathbf{x}_{long} + B_{long}\mathbf{u}_{long} \tag{4.7}$$

$$\mathbf{y}_{long} = C_{long}\mathbf{x}_{long} + D_{long}\mathbf{u}_{long}$$

where, $\mathbf{x}_{long} = \begin{bmatrix} \omega_{fl} & \omega_{fr} & V_x \end{bmatrix}^T$ denotes the states, $\mathbf{u}_{long} = \begin{bmatrix} T_{d\_fl} & T_{d\_fr} & \delta_f \end{bmatrix}^T$ denotes the control inputs and $\mathbf{y}_{long} = \begin{bmatrix} V_x & \omega_{fl} & \omega_{fr} \end{bmatrix}^T$ denotes the measured outputs.
The non-linear dynamics model is linearized at different operating points. Some of the operating points and the resultant linearized matrices are listed below.

Case 1: $V_x = 20kmph$

$$A_{long} = \begin{bmatrix} -12.81 & 0.053 & -42.61 \\ 0.053 & -12.81 & -42.61 \\ 0.11 & 0.11 & 0.6999 \end{bmatrix}, \quad B_{long} = \begin{bmatrix} 0.98 & 0 \\ 0 & 0.98 \\ 0 & 0 \end{bmatrix}$$

$$C_{long} = \begin{bmatrix} 0 & 0 & 1 \\ 1 & 0 & 0 \\ 0 & 1 & 0 \end{bmatrix}, \quad D_{long} = \begin{bmatrix} 0 & 0 \\ 0 & 0 \\ 0 & 0 \end{bmatrix}$$

Eigen values are: [-0.0078; -12.8660; -12.0529]

Case 2: $V_x = 30kmph$

$$A_{long} = \begin{bmatrix} -8.81 & 0.042 & -29.1 \\ 0.042 & -8.81 & -29.1 \\ 0.073 & 0.073 & 0.47 \end{bmatrix}, \quad B_{long} = \begin{bmatrix} 0.98 & 0 \\ 0 & 0.98 \\ 0 & 0 \end{bmatrix}$$

Eigen values are: [-0.0116; -8.8421; -8.2760]

Case 3: $V_x = 55kmph$

$$A_{long} = \begin{bmatrix} -7.03 & 0.038 & -23.08 \\ 0.038 & -7.03 & -23.08 \\ 0.058 & 0.058 & 0.37 \end{bmatrix}, \quad B_{long} = \begin{bmatrix} 0.98 & 0 \\ 0 & 0.98 \\ 0 & 0 \end{bmatrix}$$

Eigen values are: [-0.0148; -7.0700; -6.6096]

4.8 Summary

i. The overall concept of integrated controller strategy is explained in detail along with a discussion on the crucial parameters which are required to analyze the controller performances in various test scenarios.

ii. A number of different test maneuvers are used in this work to provide useful information about different aspects of the dynamic behavior of the vehicle and the effectiveness of each control strategy. These test maneuvers are chosen such that all aspects of vehicle dynamics are addressed. In other words, the chosen test maneuvers act to clarify the performance and effect of different chassis control systems on the driver-vehicle-environment control loop and quantify the advantages of each control strategy.

iii. The linearization process of the non-linear NTTW vehicle model is explained in detail. The non-linear lateral-tilt and longitudinal models are linearized separately for control design and the validation of these models are documented.

Chapter 5

Design of Baseline Controller Strategy

5.1 Introduction

As a first step towards designing a dynamic stability controller for the NTTW vehicle, it is appropriate to tackle the fundamental issue of tilt stability while maintaining the desired driver maneuver. It is also befitting to first study the tilting motion controller using PID type control strategy because these controllers are extensively employed in the industry. This controller is named as a baseline controller strategy as this will form a benchmark for the conduct of further study on advanced controllers for the control objectives elaborated in Chapter 4. This chapter describes details of the baseline controller strategy and its design. Simulation studies of the baseline controller for constant radius turn are carried out to analyze the behavior of the NTTW vehicle with baseline controller. Further, a double lane change maneuver is used to compare the baseline controller with MVTD controller which is reported in Chapter 6.

5.2 Baseline controller design

The baseline controller comprises of two schemes, one for lateral-tilt dynamics and the other for longitudinal dynamics. The driver inputs are δ_{sw} and V_{x_des}. The basic goal of the control strategy for the lateral-tilt dynamics is to track the desired tilt angle. The δ_{sw} is a direct input to the NTTW vehicle model and the reference model and it calculates the θ_{des} input to the controller. The block diagram of the baseline control strategy is shown in Figure 5.1. The controller for non-linear lateral-tilt dynamics is designed using proportional-integral-derivative (PID) technique. Two cascade controllers are implemented to generate a control input (T_{tilt}) to track the θ_{des} and rate of change of tilt angle ($\dot{\theta}$). It has been already explained in Chapter 2, that the tilting of NTTW vehicle can be achieved by two different mechanisms, i.e. steer by wire and tilt torque. Here, the

PID control technique is a single-input system, hence the tilting is provided by the tilting torque actuator only. The steering is equal to the driver steering itself. Hence, the baseline controller uses the DTC mechanism. The control objective for non-linear longitudinal dynamics is to track the V_{x_des} commanded by the driver. The driver's speed request and the actual speed of the NTTW vehicle are compared and the error is input to the PID controller. The output is control drive torque (T_{d_req}) which is divided equally into the two wheels (T_{d_fl} and T_{d_fr}) to propel the vehicle with the desired speed. All of the system states are assumed to be available for the controller.

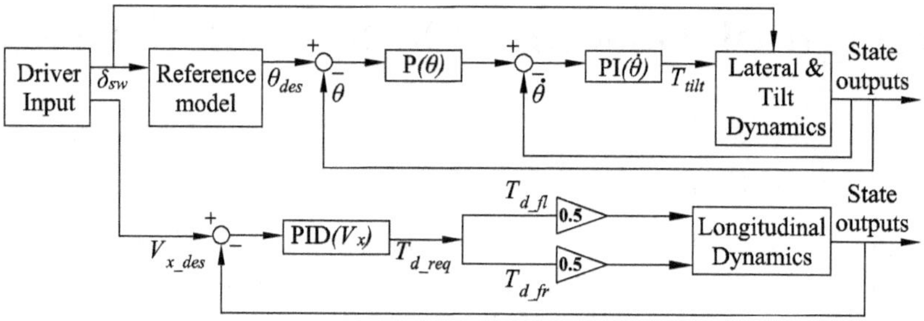

Figure 5.1: Architecture of baseline controller

The control input for the lateral-tilt dynamics is designed as:

$$u_{lat} = k_{p_2} e_{lat} + k_i \int_0^t e_{lat}(\tau) d\tau \tag{5.1}$$

where, k_{p_2}, k_i denote the proportional and integral constant terms of PI($\dot{\theta}$) controller and k_{p_1} denotes the proportional constant of P(θ) controller shown in Figure 5.1. And defining,

$$e_{lat} = k_{p_1}(\theta_{des} - \theta) - \dot{\theta}$$

The controller for longitudinal dynamics is designed as:

$$u_{long} = K_p e_{long} + K_i \int_0^t e_{long}(\tau) d\tau + K_d \frac{de_{long}}{dt} \tag{5.2}$$

where, K_p, K_i and K_d denote the proportional, integral and derivative constant terms of PID(V_x) controller shown in Figure 5.1.

$$e_{long} = V_{x\_des} - V_x$$

The PID gains of the baseline controller are tuned through trial and error method. The gain values of proportional (k_p), integral (k_i) and derivative (k_d) terms and are shown in Table 5.1.

Table 5.1: Controller gains of baseline controller

Gain parameter	Control gain values			
	P(θ)	PI($\dot{\theta}$)	PID(V_x=20 kmph)	PID(V_x=30 kmph)
k_p	5	33	400	300
k_i	-	5	10	10
k_d	-	-	10	30

It is observed that the PID gain values of the longitudinal controller need to be modified for different values of speed. In addition, the baseline controller will require extensive gain tuning for complex maneuvers such as cornering while acceleration, which is one of the test maneuvers considered in this research work.

5.3 Results and Discussion

This section presents the simulation results of the baseline controller for constant radius turn maneuver. Initially, the vehicle moves in a straight line for 5 sec. Then, the driver takes a left turn of 35 m radius of curvature at 30 kmph ($\approx 8ms^{-1}$) and continues with this constant radius until the end of the simulation. Figure 5.2 through Figure 5.6 show plots for the control inputs, the vehicle lateral and longitudinal states, body accelerations and the vehicle trajectory.
The driver steering angle (δ_{sw}) is directly fed to the vehicle steering system ($\delta_{sw} = \delta_f$), whereas the tilting torque (T_{tilt}) and drive torques (T_d) are the baseline controller inputs as shown in Figure 5.2. The NTTW vehicle lateral-tilt and longitudinal states for baseline controller are shown in Figure 5.3 and Figure 5.4 respectively. It can be observed from Figure 5.3 that the controller is able to track the desired tilt angle and the tilt rate at steady state is zero which is the desired behavior. However, the desired yaw rate is not achieved which results in deviation of the NTTW vehicle trajectory from the desired path as shown in Figure 5.6.
Similarly, the longitudinal states are plotted in Figure 5.4. The desired speed is tracked successfully by the baseline controller. Even though the angular speeds of the longitudinal dynamics are not controlled by the baseline controller, those states are monitored to analyze their responses in open loop. The steady state value of the angular velocities should be 20 $rads^{-1}$. The steady state values match with the expected values, however, the transient peak is quite high which requires minimization for smoother operation. The NTTW vehicle body accelerations are longitudinal (a_x), lateral (a_y) and perceived lateral (a_{per}) accelerations. The steady state value of the perceived lateral acceleration is zero denoting lateral stability of the NTTW vehicle as mentioned in Section

4.3.1. The value of a_y is also within the specified limit (explained in Chapter 4).

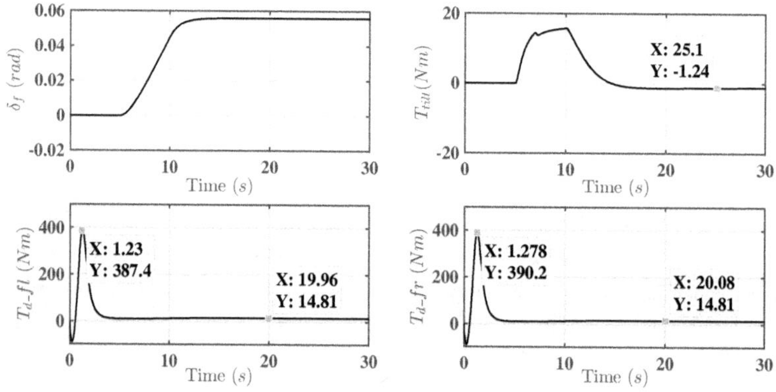

Figure 5.2: Inputs for constant radius turn

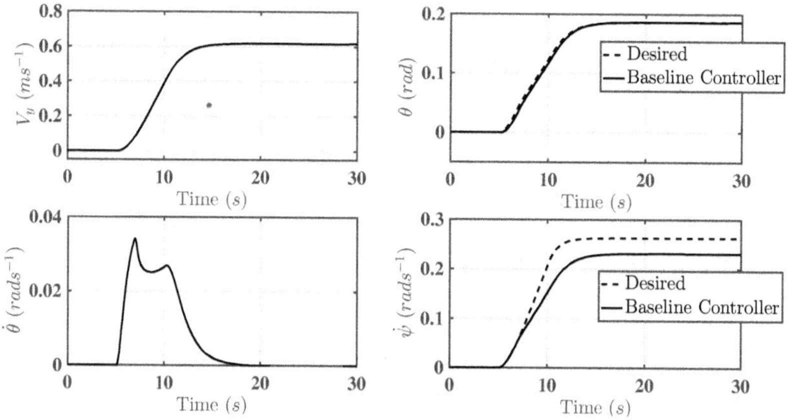

Figure 5.3: Lateral states for constant radius turn

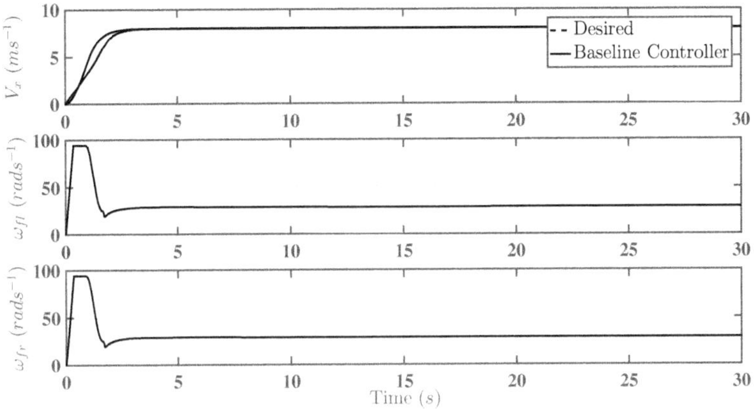

Figure 5.4: Longitudinal states for constant radius turn

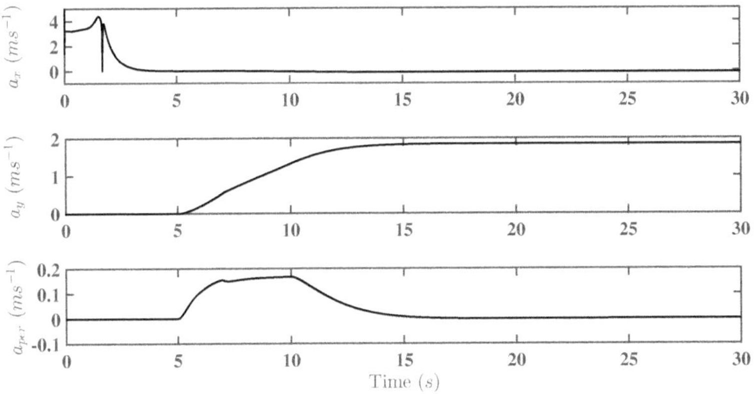

Figure 5.5: Body accelerations for constant radius turn

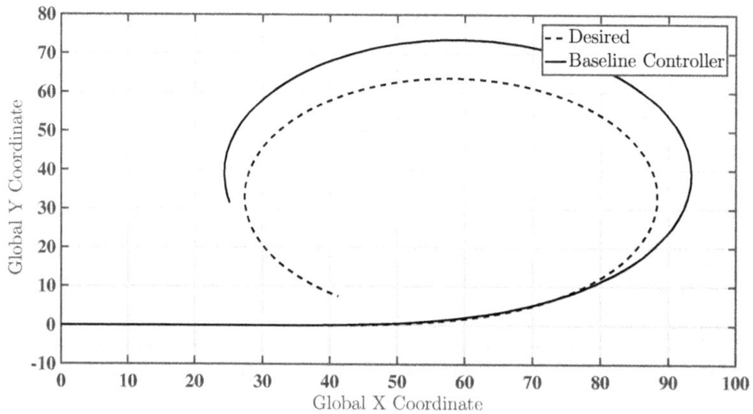

Figure 5.6: The trajectory of NTTW vehicle for constant radius turn

It can be inferred from the above studies that the objectives of tilt stability of the NTTW vehicle along with minimum steady state T_{tilt} and zero steady-state value of a_{per} is met using baseline controller. But, the overall response of the baseline controller shows that the desired trajectory is not followed satisfactorily due to the steady-state error present in the other lateral state, i.e. the yaw rate. It can be concluded that a control strategy is required which can simultaneously achieve the reference values of the yaw rate state as well as the tilt states of the vehicle while maintaining the desired longitudinal speed, which would essentially mean that the driver's command is being followed. In addition, it is observed that the control gains for longitudinal dynamics controller need to be changed for different velocities to maintain desired tracking (V_x). Hence, there is a need to apply an advanced control strategy for longitudinal dynamics to achieve vehicle speed tracking across a wider envelope of operation with a single gain matrix. Further simulation studies of the baseline controller in presence of external disturbances are reported during performance comparison with the proposed MVTD controller in Chapter 6.

5.4 Summary

i. In this chapter, the design of PID based baseline controller has been reported. The baseline controller deals with the primary objective of the active tilt control system of the NTTW vehicle, i.e. tracking of desired tilt angle along with the desired longitudinal velocity. The control inputs are the required tilting torque (T_{tilt}) and the required drive torque(T_d). The driver steering angle depicting the required yaw motion is directly fed to the vehicle. This essentially illustrates that baseline controller is based on DTC system.

ii. It can be observed from the obtained results that the baseline controller is able to maintain lateral and longitudinal stability for a constant radius of turn maneuver. However, it is not able to track the desired trajectory satisfactorily. In addition, for different longitudinal speeds, the PID gains need to be tuned and it needs further complex tuning for maneuvers such as accelerating while cornering. Hence, a more advanced controller needs to be designed to efficiently cover a larger operating range with a single gain matrix and to simultaneously achieve the reference values of the yaw rate state as well as the tilt states of the vehicle while maintaining stability.

Chapter 6

Design of Multivariable Tilt and Drive (MVTD) Controller Strategy

6.1 Introduction

It is evident from the results obtained in Chapter 5 that there is a need for a technique that accounts for the control of multiple states of the NTTW vehicle simultaneously. Also, this necessitates usage of multiple inputs which would provide the designer with the required flexibility and consequently, avoid the problem of underactuation. To accomplish such a MIMO (multiple input multiple output) system control, one of the popular state space methods is state feedback technique. The concept of feeding back all the state variables to the input of the system through a suitable feedback matrix in the control strategy is known as the full-state variable feedback control technique [66]. In this chapter, Linear Quadratic Regulator (LQR) based MIMO controllers have been designed for lateral-tilt and longitudinal dynamics and then extensively refined to achieve the desired performance. This optimal control technique is designed for the integrated dynamics stability control of the NTTW vehicle. The term 'integrated' signifies a simultaneous control approach for the lateral-tilt and longitudinal dynamics as explained in Chapter 4.

The theory of the LQR control strategy is first explained in brief and then the design of Multivariable Tilt and Drive (MVTD) controller strategy is explained in detail in this chapter. Extensive simulation studies are performed using chosen test maneuvers for the performance analysis of the NTTW vehicle using MVTD controller. Additionally, comparison with baseline controller is reported to analyze the efficacy of the controllers.

6.2 Background on optimal control algorithm

The linear quadratic regulator (LQR) is a well-known state feedback design technique that is used to derive control gains. One of the major rationales towards seeking an optimum controller is that optimal response of the system can be obtained in accordance with some specifications, for example, the actuator limits [67].

6.2.1 Formulation of LQR

In this section, a generic formulation of LQR control strategy is explained in brief. System dynamics of a plant under consideration can be represented in state space form as:

$$\dot{\mathbf{x}} = A\mathbf{x} + B\mathbf{u}$$

$$\mathbf{y} = C\mathbf{x} + D\mathbf{u} \tag{6.1}$$

It is assumed that all the states '**x**' are accessible to the controller. The feedback gain matrix 'K_{lqr}' for the optimal control vector '**u**' is represented by equation (6.2).

$$\mathbf{u} = -K_{lqr}(\mathbf{x} - \mathbf{x}_{des}) \tag{6.2}$$

The optimal control vector '**u**' is designed such that a performance criterion is met, to be specific, **u** is obtained to minimize the error between the desired and actual states of the system in order to track the reference trajectory. The LQR strategy minimizes a quadratic cost function J to find out the gain K_{lqr} [68].

$$J = \int ((\mathbf{x} - \mathbf{x}_{des})^T Q(\mathbf{x} - \mathbf{x}_{des}) + \mathbf{u}^T R\mathbf{u})dt \tag{6.3}$$

Here, Q is a real symmetric positive semi-definite matrix and identifies the relative significance of the error between desired and actual response of the states. The matrix R is real symmetric positive definite which ascertains the energy expenditure by control signals. The matrix Q and R tuned through trial and error method to minimize the steady-state error of system states.

6.2.2 Addition of Integral action in LQR loop

An integral action is added to the LQR framework in order to not only restrict the study to null stabilization but also to ensure zero steady-state error while tracking. The idea is to consider an external control loop and stabilize the integral of the error between the measurement and the desired set point to track [69]. This controller is termed as LQR with integral action (LQI). The relevant state errors are integrated and fed back to the controller to achieve zero steady-state error.

6.3 Design of MVTD controller

An integrated multivariable control strategy termed as multivariable tilt and drive (MVTD) controller is designed using optimal control strategy. The integrated controller consists of two separate control schemes, one for nonlinear lateral-tilt and the other for longitudinal dynamics. The prime objective is to achieve the desired trajectory of the driver while maintaining vehicle stability. The control inputs are the modified steering angle (steer-by-wire actuator) and the tilting torque (tiling actuator) which renders the strategy to be based on SDTC system. Figure 6.1 shows the overall block diagram of the MVTD control architecture. The nonlinear lateral-tilt dynamics are controlled using LQI control strategy and LQR control strategy is used for nonlinear longitudinal dynamics and their respective design procedures are given in subsequent sub-sections. The control inputs of the integrated controller are the δ_f, T_{tilt} for the lateral-tilt dynamics and T_{d_i} (drive torque of the ith wheel) for the longitudinal dynamics. It is assumed that all the states are available for the controller.

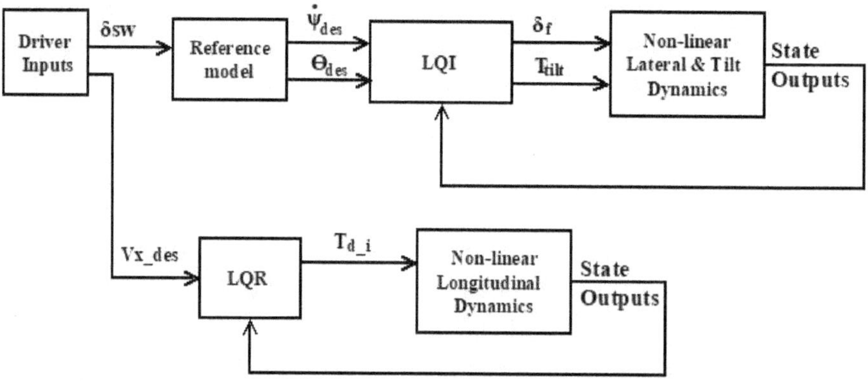

Figure 6.1: Architecture of MVTD controller

6.3.1 Design of control strategy for lateral-tilt dynamics

The controller for the non-linear lateral-tilt dynamics is designed using linearized system matrices as it is more beneficial for control design and practical implementation, especially when deploying the controller in embedded systems. The linearized system matrices denoted by equation (4.6), as explained in Chapter 4, for lateral-tilt control design are given below:

$$A_{lat} = \begin{bmatrix} -28.2 & 0 & 31.2 & -8 \\ 27.8 & 0 & -21.1 & -9.21 \\ 0 & 1 & 0 & 0 \\ 0.08 & 0 & 2.76 & -14.1 \end{bmatrix}, \quad B_{lat} = \begin{bmatrix} 147.7 & -0.0053 \\ -145.8 & 0.0088 \\ 0 & 0 \\ 66.3 & 0 \end{bmatrix}$$

$$C_{lat} = \begin{bmatrix} 1 & 0 & 0 & 0 \\ 0 & 0 & 0 & 1 \\ 0 & 1 & 0 & 0 \\ 0 & 0 & 1 & 0 \end{bmatrix}, \quad D_{lat} = \begin{bmatrix} 0 & 0 \\ 0 & 0 \\ 0 & 0 \\ 0 & 0 \end{bmatrix}$$

The eigenvalues for V_x = 30 kmph are: [-27.02; 2.6; -3.8; -14.1]. It can be observed that the system is inherently unstable due to the presence of an unstable pole.

As explained earlier, the lateral-tilt dynamics is controlled using a linear quadratic regulator with integral action (LQI). The LQR is a proportional state feedback control method which means that the state tracking is not offset free. An integral action is required to minimize the steady-state error. It is observed that there is a need to minimize the steady-state error in the tracking of θ and $\dot{\psi}$ to achieve the desired performance. Hence, two integral error states are augmented in the system matrices. The system matrix for the lateral-tilt dynamics controller is hence modified ($[A_{lat}, B_{lat}, \mathbf{x}_{lat} \implies A'_{lat}, B'_{lat}, \mathbf{x}'_{lat}]$) by augmenting state vector with two integral error states (θ and $\dot{\psi}$). The lateral-tilt controller is designed as:

$$\mathbf{u}_{lat} = -k_{lat} e_{lat} \tag{6.4}$$

Defining,

$$e_{lat} = (\mathbf{x}'_{lat} - \mathbf{x}'_{lat\_des})$$

where, e_{lat} are the lateral-tilt dynamics state error signals which are obtained by subtracting actual states (V_y,$\dot{\theta}$,θ and $\dot{\psi}$) from desired states (as obtained in Section 4.3.1). The desired values of $\dot{\theta}$ and integral states are zero. The pattern of the V_y state is expected to have a correlation with the steering wheel angle of the vehicle. The tracking of V_y is not a part of the control objectives of this work and hence is not addressed further. The state feedback control gain values for lateral-tilt dynamics (k_{lat}) used in the MVTD controller is given in Table 6.1. k_{lat} in Table 6.1 denotes the LQI control gains for each element of the state vector of lateral-tilt dynamics.

Table 6.1: State feedback control gain values for lateral-tilt dynamics in MVTD controller

Controller		k_{V_y}	$k_{e_{\dot{\theta}}}$	k_{e_θ}	$k_{e_{\dot{\psi}}}$	$k_{\int e_\theta}$	$k_{\int e_{\dot{\psi}}}$
LQI lateral gains	δ_f	0.009	−0.018	−0.05	0.6	−0.09	0.52
	T_{tilt}	6.95×10^2	14.5×10^2	63.2×10^2	-2.1×10^2	43.3×10^2	11.6×10^2

6.3.2 Design of control strategy for longitudinal dynamics

In the baseline controller discussed in Chapter 5, a PID controller is implemented for the longitudinal dynamics to achieve V_{x_des}. However, it is observed that the PID gains need to be changed for different velocities to maintain the desired tracking. Hence, there is a need to apply an advanced control strategy for longitudinal dynamics of the NTTW vehicle. The aim is to achieve vehicle speed tracking across a wider envelope of operation. To achieve this, an LQR controller is designed for the longitudinal vehicle dynamics. Such a controller involves a single state feedback gain across a range of vehicle speeds. Similar to lateral-tilt controller design, the longitudinal dynamics controller is also designed using linearized system matrices. The linearized system matrices denoted by equation (4.7) as explained in Chapter 4 for longitudinal dynamics control design are given below:

$$A_{long} = \begin{bmatrix} -12.81 & 0.053 & -42.61 \\ 0.053 & -12.81 & -42.61 \\ 0.11 & 0.11 & 0.6999 \end{bmatrix}, \quad B_{long} = \begin{bmatrix} 0.98 & 0 \\ 0 & 0.98 \\ 0 & 0 \end{bmatrix}$$

$$C_{long} = \begin{bmatrix} 0 & 0 & 1 \\ 1 & 0 & 0 \\ 0 & 1 & 0 \end{bmatrix}, \quad D_{long} = \begin{bmatrix} 0 & 0 \\ 0 & 0 \\ 0 & 0 \end{bmatrix}$$

The eigenvalues for $V_x = 30$ kmph are: [-0.0078; -12.8660; -12.0529]. Thus, the dynamics are inherently stable. The LQR gains are obtained by tuning Q and R through a trial and error method. The longitudinal control input is defined as:

$$\mathbf{u}_{long} = -k_{long} e_{long} \tag{6.5}$$

Defining,

$$e_{long} = (\mathbf{x}_{long} - \mathbf{x}_{lat\_long})$$

where, e_{long} are the error signals of longitudinal dynamics, which are obtained by subtracting actual states from desired states (as obtained in Section 4.3.2). The state feedback control gain (k_{long}) values for longitudinal dynamics used in the MVTD controller is given in Table 6.2. k_{long} in Table 6.2 denotes the LQR control gains for each element of the state vector of longitudinal dynamics.

Table 6.2: State feedback control gain values for longitudinal dynamics in MVTD controller

Controller		$k_{e_{\omega_{fl}}}$	$k_{e_{\omega_{fr}}}$	$k_{e_{V_x}}$
LQR longitudinal gains	T_{fl}	2.3×10^2	0.6	15.73×10^2
	T_{fr}	0.6	2.3×10^2	15.73×10^2

6.4 Results and Discussion

In this section, a detailed simulation analysis has been documented for the performance of the MVTD controller for the selected maneuvers as discussed in Section 4.5. To start with, the performance of baseline controller, as discussed in Chapter 5, has been compared with that of MVTD controller to understand the efficacy and usefulness of the MVTD controller over the baseline controller or vice-versa. In the subsequent sub-sections, a detailed analysis of the MVTD controller is presented through extensive simulations. The test maneuvers discussed in Table 4.2 are used in the analysis throughout the thesis work to maintain uniformity. The baseline and MVTD controllers are simulated at 2 ms.

6.4.1 Comparison with baseline controller

A double lane change maneuver with around 23 m of turning radius and constant speed of 20 kmph ($\approx 8ms^{-1}$) is simulated to compare the two control strategies proposed in Chapter 5 and Section 6.3. Initially, the vehicle moves in a straight line for 5 sec. Then, the driver takes a turn of 23 m radius of curvature. At 20 seconds, the driver changes his direction of travel to a straight line till the end of the simulation. Figure 6.2 through Figure 6.5 show comparison plots for the control inputs, the controlled states and the trajectories of the NTTW vehicle model. Further comparison studies cover key parameters such as perceived lateral acceleration, robustness analysis and power requirements of control inputs.

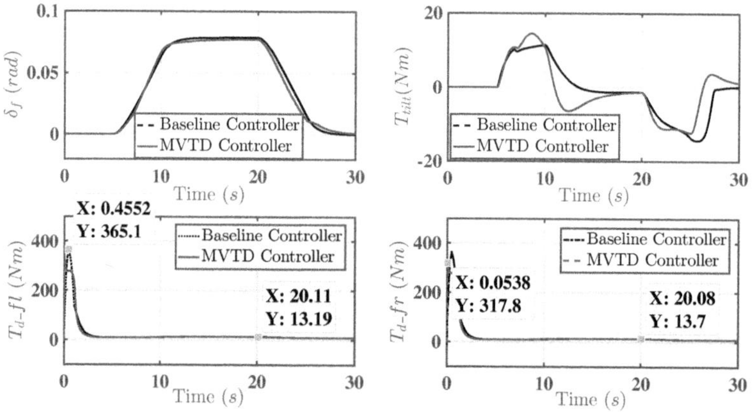

Figure 6.2: Control inputs of the two strategies

The control inputs required for the desired maneuver are shown in Figure 6.2. The steering input for lateral-tilt dynamics control and torque input for longitudinal dynamics control for both the strategies are almost same. It is also observed during steady state, the inputs of longitudinal dynamics for both the control strategies are similar. However, the peak drive torque values are lower in case of MVTD controller ($\approx$ 318 Nm) than the baseline controller (365 Nm).

Figure 6.3 denotes the plots for tilt, yaw rate and longitudinal velocity tracking. The 'Desired' signals shown in the plots are generated from the reference model. The steady-state error of yaw rate tracking is non zero for baseline controller and hence the calculated desired signal ('Desired New') for tilt angle is different (lower) than the one ('Desired') calculated for the MVTD controller as given in equations (4.2) through (4.5). On the other hand, yaw rate tracking has zero steady state error and hence the optimal desired tilt angle is generated in case of MVTD controller. However, in both the strategies, the tilt controller is performing satisfactorily. In case of longitudinal dynamics control, both the strategies seem to work well at a particular speed. It should be noted that for the baseline controller strategy, the PID gains are needed to be re-tuned for different desired speed requirements.

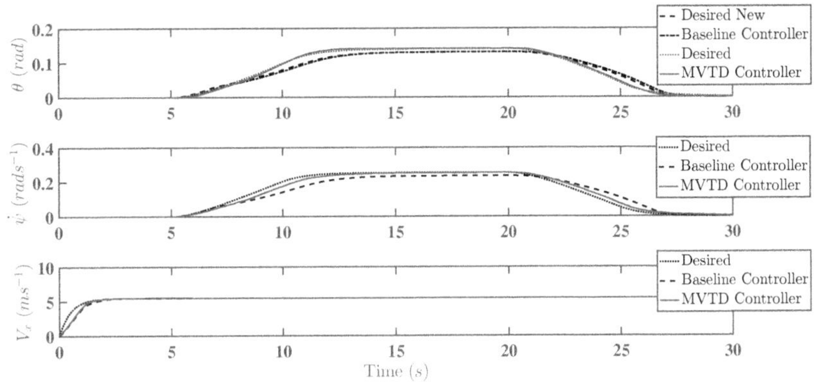

Figure 6.3: Dynamic states of the two strategies

The perceived lateral acceleration tends to zero when the desired tilt angle is achieved by both the baseline as well as MVTD controllers as shown in Figure 6.4. This signifies that the stability of lateral dynamics of the NTTW vehicle is achieved satisfactorily by both control strategies.

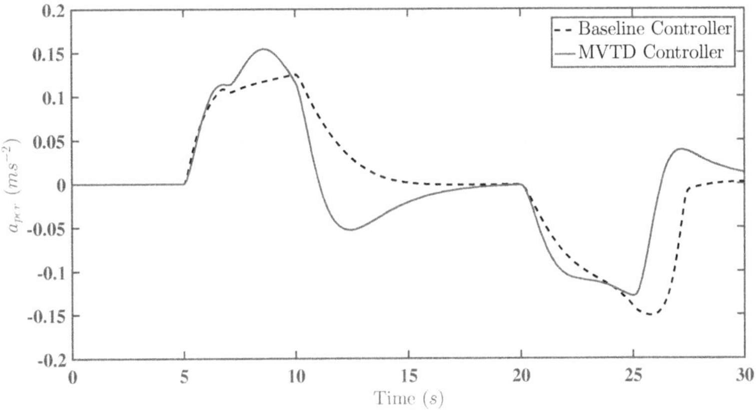

Figure 6.4: Perceived lateral accelerations of the two strategies.

The actual versus desired trajectories of baseline and MVTD controllers are plotted in Figure 6.5. It is evident from the plot that the desired trajectory is followed with less error by MVTD controller. The baseline controller is not able to follow the desired trajectory as well as the desired yaw rate is not achieved. Thus, in order to achieve the driver's desired path, the desired yaw rate must be achieved with less error.

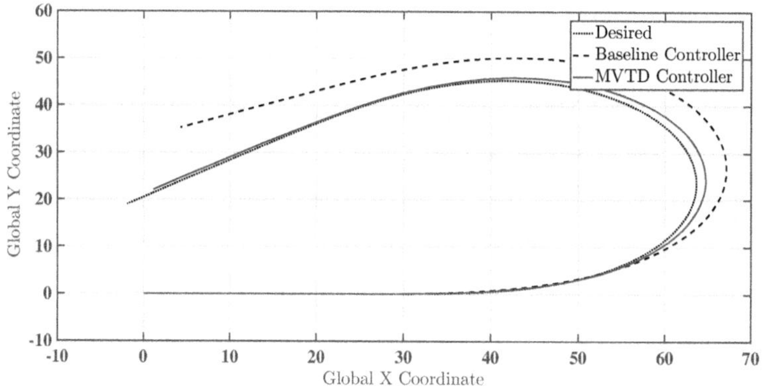

Figure 6.5: XY trajectories of the two strategies

Figure 6.6 and Figure 6.7 show plots of yaw rate state and trajectory of the NTTW vehicle model when an external sinusoidal disturbance is added to the longitudinal drag force of the vehicle at 14 s. A 5% of the maximum tyre force is added as a disturbance to longitudinal drag force. It should be noted that during a dynamic maneuver, the same disturbance is also coupled to the lateral dynamics due to the influence of tyre model. A step-steer maneuver is simulated to evaluate the performance. Initially, the vehicle moves in a straight line for 5 sec followed by a turn of 23 m radius of curvature. The speed of the vehicle is kept constant at 20 kmph ($\approx 5.5ms^{-1}$). The results in Figure 6.6 and Figure 6.7 indicate the coupling within the lateral and longitudinal dynamics of the NTTW vehicle model. Both the dynamics are interlinked through tyre model and hence change in longitudinal will affect the lateral dynamics as well. In Figure 6.6 this coupling takes place after 14 second when step-steer is commanded due to the interaction between longitudinal and lateral-tilt dynamics from tyre model. Figure 6.7 indicates that the addition of disturbance renders the NTTW vehicle towards instability in case of baseline controller. Whereas, the MVTD controller is performing satisfactorily under the same condition.

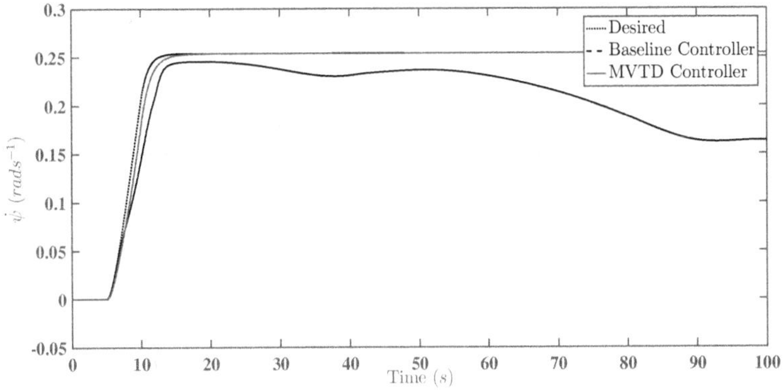

Figure 6.6: Tracking of yaw rates of the two strategies with disturbance

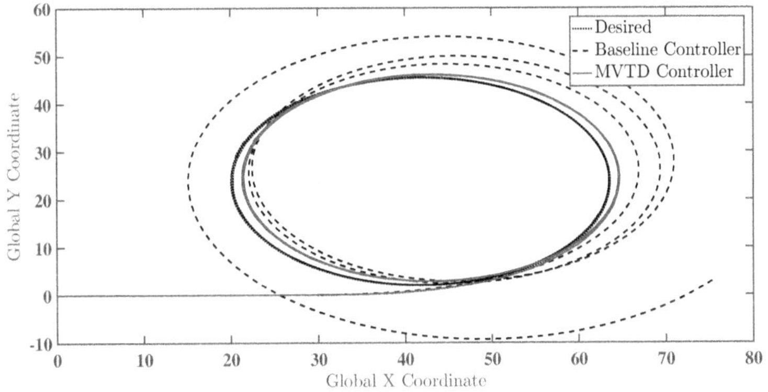

Figure 6.7: XY trajectories of the two strategies with disturbance

To test the robustness of the control strategy, the system dynamics parameters, such as the mass of the NTTW vehicle is varied by 20%. The tilt angle tracking is plotted to check the robustness of the MVTD controller. Figure 6.8 and Figure 6.9 show the plots of tilt angle tracking when system parameters are varied. It can be observed that the MVTD controller is more robust to variation in mass than baseline controller.

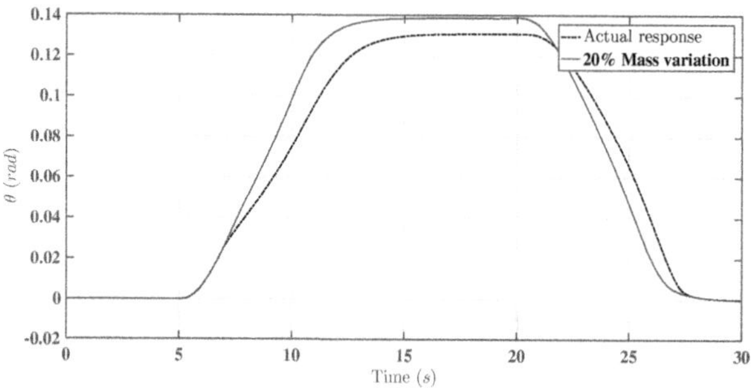

Figure 6.8: Effect of parameter variation in tracking of the tilt angle for Baseline controller

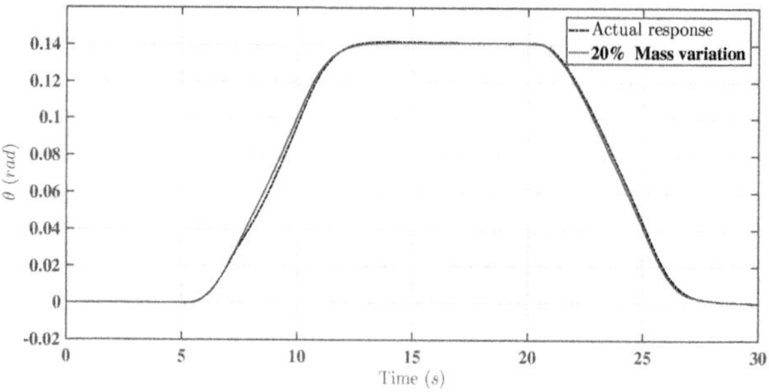

Figure 6.9: Effect of parameter variation in tracking of the tilt angle for MVTD controller

Input power calculation

To evaluate the actual difference between the control efforts ($\mathbf{u}$) of the MVTD controller and baseline controller in quantitative terms, power utilization of the control efforts is calculated. The mathematical expression for the equivalent power is given by (6.6). Table 6.3 provides the total input power of each of the control strategies during two different test maneuvers. It can be observed that the baseline controller strategy utilizes more power than MVTD controller in both the test scenarios.

$$Power = \frac{1}{length(\mathbf{u})}(\mathbf{u}^2) \tag{6.6}$$

Table 6.3: Total input power of the two strategies

Test case	P(MVTD)	P(Baseline)	Difference
30 kmph and 32 m turn radius	5.7227e+03 W	8.5759e+03 W	50 %
20 kmph and 23 m turn radius	5.7227e+03 W	7.4060e+03 W	25 %

Remark 1. It is observed that tracking of longitudinal velocity and tilt angle are achieved through baseline and MVTD controllers satisfactorily. Whereas, the yaw rate and trajectory tracking are better for MVTD controller. It can also be inferred from the simulations that the MVTD controller is less sensitive to external disturbances and is more robust to parametric variation. In addition, it can be inferred that the MVTD control strategy utilizes lower input power than the baseline controller which is a desired performance characteristic. During the simulation studies, it is also observed that the operating envelope of the MVTD controller is larger (in terms of minimum turning radius) than that of the baseline controller. The results of simulations indicate that the proposed novel MVTD strategy provides better vehicle handling and stability performance.

6.4.2 Double lane change maneuver

This section presents the simulation results of the MVTD controller for various test scenarios. The main aim is to analyze the responses of the dynamic variables of the NTTW vehicle, such as tyre slip angles, tyre forces etc., while various degrees of cornering. The first case of simulation is performed to show the complete set of dynamic responses of the NTTW vehicle during a double lane change maneuver. Initially, the vehicle moves in a straight line for 5 sec. Then, the driver takes a left turn of 10 m radius of curvature at a constant speed of 10 kmph ($\approx 2.8ms^{-1}$). At 22 seconds, the driver changes his direction of travel to a straight line till the end of the simulation. Figure 6.10 through Figure 6.19 show plots for the control inputs, the vehicle states, tyre forces, tyre slip angles, and the body accelerations. This maneuver is quite challenging as compared to the ones provided in the literature.

The δ_{sw} is modified to δ_f by 1.8% to attain the driver's desired trajectory while the vehicle is tilting appropriately to achieve stability. The tilting torque transiently increases then settles to around 2 Nm as steady state is achieved. The peak drive motor torques for front and right wheels are 288 Nm and average torque is 12.4 Nm during cornering. However, as the vehicle travels in a straight path the average torque is around 8.2 Nm.

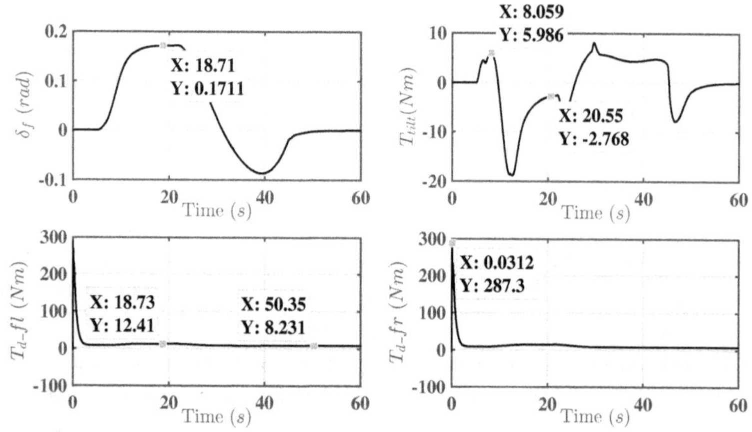

Figure 6.10: Control inputs for double lane change maneuver

All the states (lateral-tilt and longitudinal dynamics) of the NTTW vehicle are plotted in Figure 6.11 and Figure 6.12. The lateral states θ and $\dot{\psi}$ are tracked against their respective desired values. The other lateral states $\dot{\theta}$ is expected to approach to zero at steady state. V_y is expected to be stable and bounded at steady state and a good correlation exists between the behavior of V_y and δ_f of the vehicle. Similarly, all the states of the longitudinal dynamics are tracked against their respective desired signals. It can be observed that the desired states are being tracked successfully. The peak transient values of the longitudinal states (ω_{fl} and ω_{fr}) are minimal which denotes a smooth operation. Figure 6.13 shows the plot of the lateral tyre forces of the NTTW vehicle while cornering. The load transfer occurs due to the dynamic maneuver, wherein it can be observed that the load is shifted to front right tyre due to which the load at front left tyre decreases. Due to acceleration, the load transfer is at the rear tyre as evident in the plot.

The front left and right longitudinal tyre forces are shown in Figure 6.14, the rear tyre force is a purely rolling tyre and hence the force is zero. The plot, Figure 6.15 showcases one of the important responses of the NTTW vehicle dynamics while cornering. The vertical forces represent road-holding capability of the vehicle. As the vehicle starts cornering in the left side, the load shifts towards the left tyre, then at steady state the load is more at the right side than at the left side.

The tyre slip angles of the front left, right and rear wheels are shown in the Figure 6.16. The ideal values should be as low as possible than the actual steering angle. Similarly, Figure 6.17 shows the plots of the slip ratios of the front left and right wheels of the NTTW vehicle during the test maneuver.

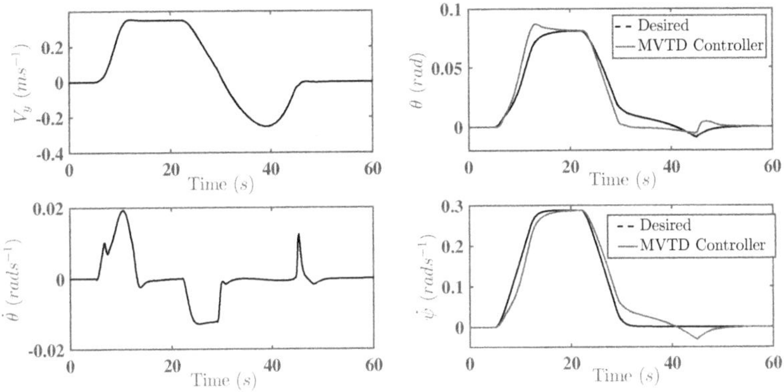

Figure 6.11: Lateral states for double lane change maneuver

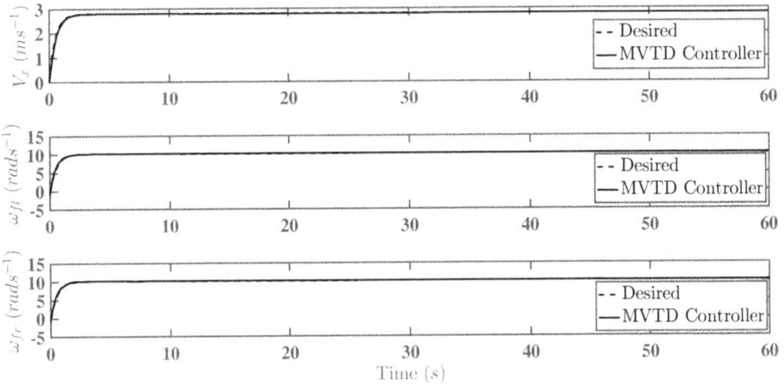

Figure 6.12: Longitudinal states for double lane change maneuver

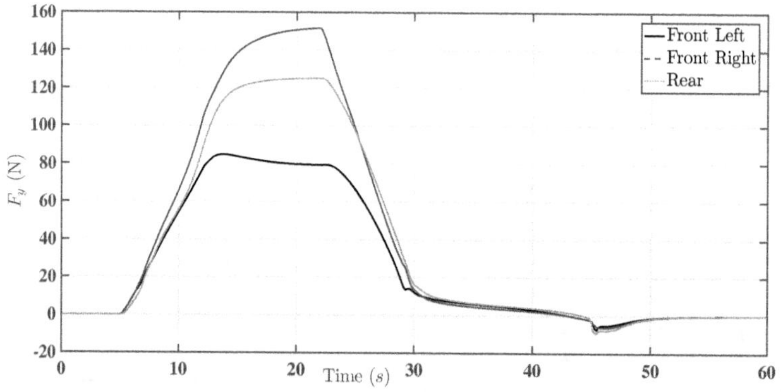

Figure 6.13: Lateral tyre forces for double lane change maneuver

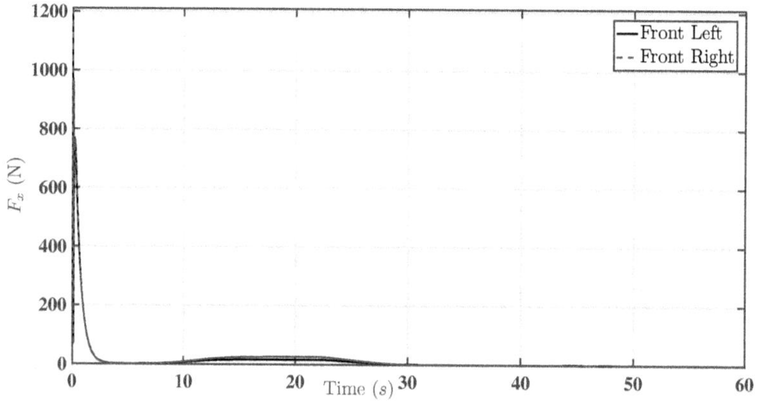

Figure 6.14: Longitudinal tyre forces for double lane change maneuver

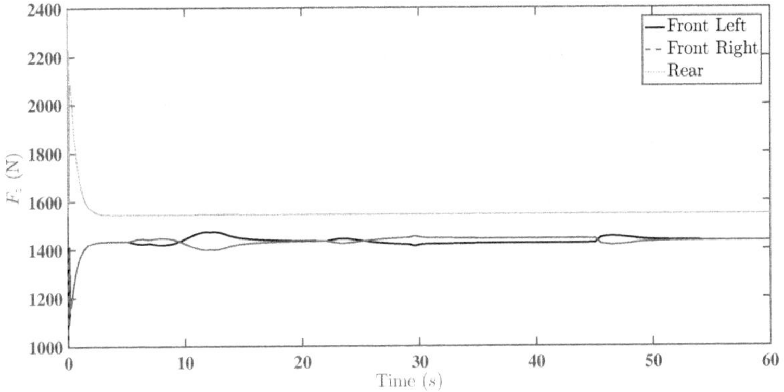

Figure 6.15: Vertical tyre forces for double lane change maneuver

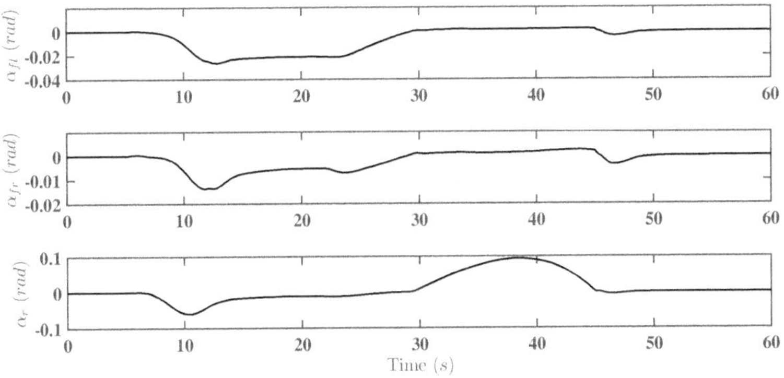

Figure 6.16: Tyre slip angle for double lane change maneuver

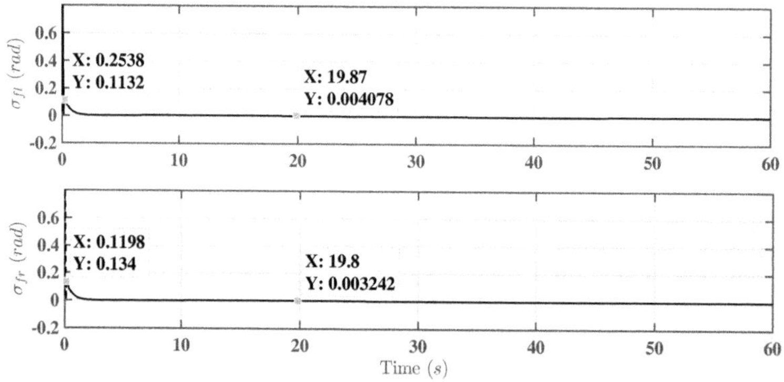

Figure 6.17: Tyre slip ratio for double lane change maneuver

The plots of longitudinal, lateral and perceived accelerations are shown in the Figure 6.18. The integrated (lateral-tilt and longitudinal) desired path is compared with the actual path in Figure 6.19. It can be observed that as the desired trajectory becomes steeper, the performance of the MVTD controller deteriorates. The trajectory tracking for such a maneuver has more error than the ones performed in the previous sections. As an overall observation, it should be noted that the settling time of the response is also high for this steep maneuver and hence needs improvement.

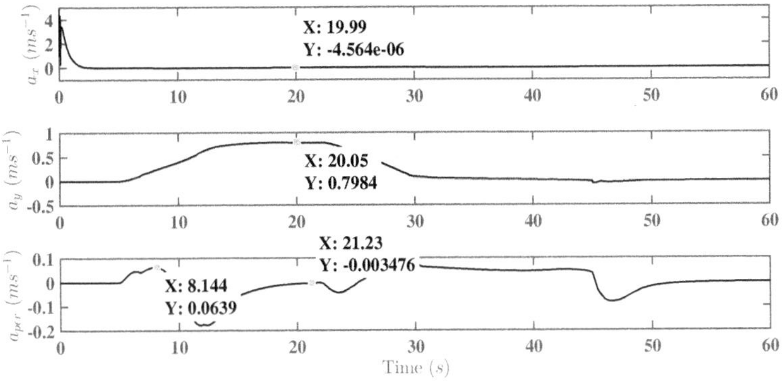

Figure 6.18: Body accelerations for double lane change maneuver

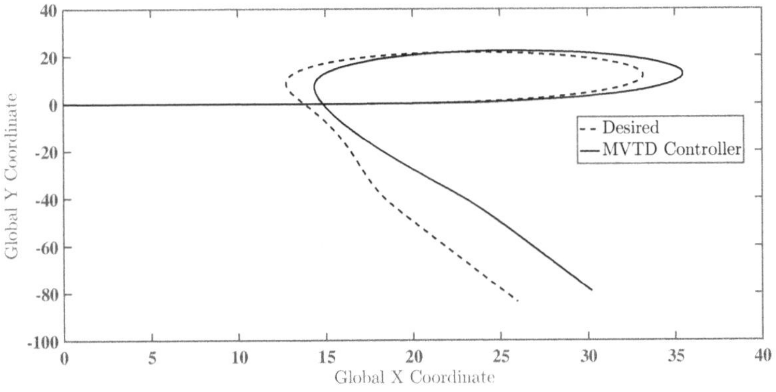

Figure 6.19: Vehicle trajectory for double lane change maneuver

6.4.3 Comparison of different radii of turns with fixed longitudinal velocity

In this section, a comparison of the performance of MVTD controller is presented through simulation studies when the driver commands different radii of turns for a fixed velocity. Figure 6.20 through Figure 6.23 provide the plots of control inputs, tyre slip angles, left tyre forces and NTTW vehicle body accelerations for 35 m, 75 m and 150 m constant turn radius at 30 kmph of vehicle longitudinal velocity. It can be observed from the plots that as the turning radius increases from 35 m to 150 m, the steering angle decreases from 0.056 rad to 0.018 rad. The required tilt torque also decreases and the steady state tilt torque moves closer to zero Nm. The plots of slip angles and the tyre forces also show a similar pattern with increasing radius of turn. These observations are in line with the understanding of dynamics of the NTTW vehicle.

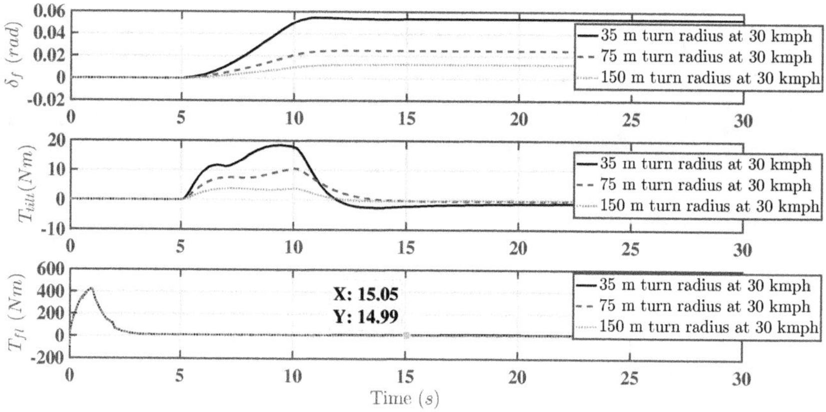

Figure 6.20: Comparison of inputs for various turns

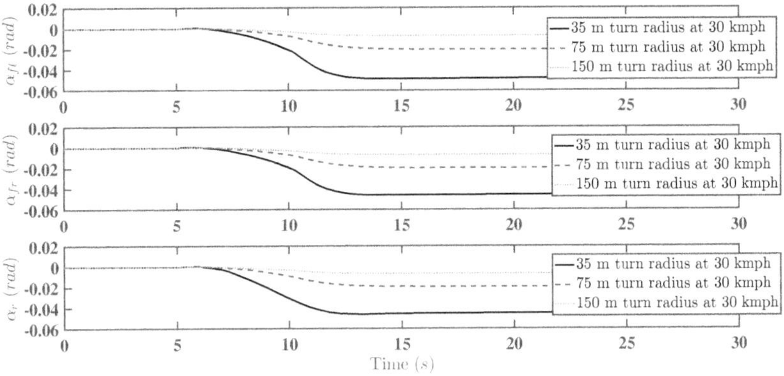

Figure 6.21: Comparison of tyre slip angles for various turns

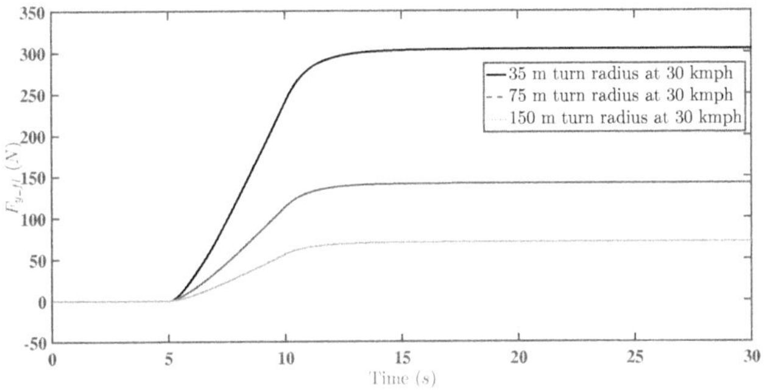

Figure 6.22: Comparison of left lateral tyre forces for various turns

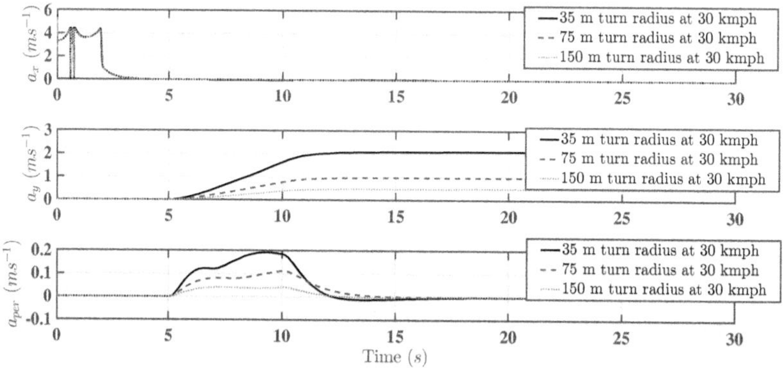

Figure 6.23: Comparison of accelerations for various turns

6.4.4 Cornering while acceleration

A three wheeled vehicle is less stable as compared to a four wheeled vehicle. [5] analyzed rollover stability while acceleration/deceleration for three wheeled vehicle, which has a larger overturning moment and consequently, lower rollover speed limits. Hence, these two tests are very important for evaluating the stability performance of NTTW vehicle. For investigation of controller behavior during acceleration/deceleration, while cornering, it is important to differentiate between large and small magnitudes. In this work, low deceleration rate (0.003 g) and moderate acceleration rate (0.1 g) are considered during simulations. A challenging scenario involves vehicle maneuvering at varying longitudinal speed while cornering. The driver (speed command) accelerates from 0 to 20 kmph ($\approx 5.5ms^{-1}$) to 75 kmph ($\approx 21ms^{-1}$) while taking a step-steer turn of fixed turning radius of curvature of 170 m. Figure 6.24 through Figure 6.29 show plots for the control inputs, the vehicle states, vertical tyre forces, body accelerations and vehicle trajectory. It can be observed from Figure 6.24 that during acceleration (20 to 35 s), the required control inputs increase and then settle when the constant steady state is achieved. Figure 6.25 represents the plot for all the lateral states. As expected, the vehicle tilts more as the speed increases from 0.01 rad (1 deg) at 20 kmph to 0.2 rad (18 deg) at 75 kmph. Further, during the acceleration time from 20 sec to 35 sec, the tilting torque increases (from 1 Nm to 15 Nm) and then settles as a steady state is achieved. It is interesting to note that the drive torque 30 Nm at 75 kmph is matching with the analytical calculations in (A.13) for the required motor torques given in Appendix A. Also, a comparison between drive torques is shown in Table A.2.

Figure 6.25 represents the plot for all the lateral states. As expected, the vehicle tilts more as the speed increases from 0.01 rad (1 deg) at 20 kmph to 0.32 rad (18 deg) at 75 kmph. The longitudinal states are tracked successfully as can be seen from the Figure 6.26.

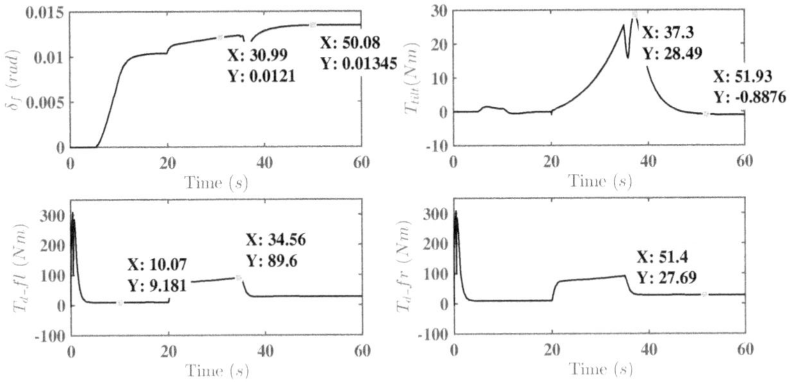

Figure 6.24: Control inputs for cornering while acceleration

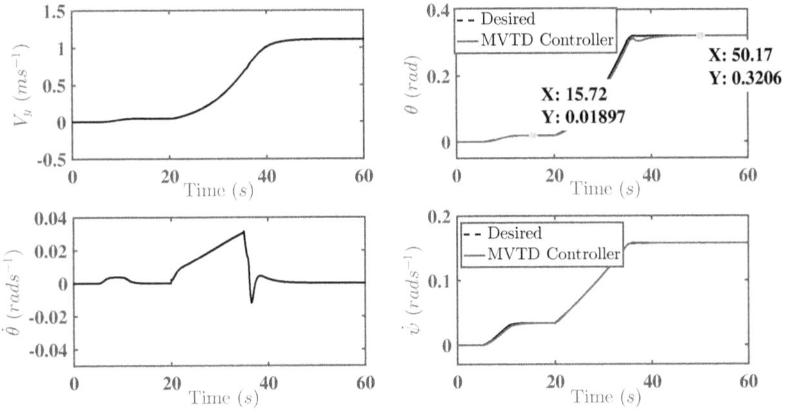

Figure 6.25: Lateral states for cornering while acceleration

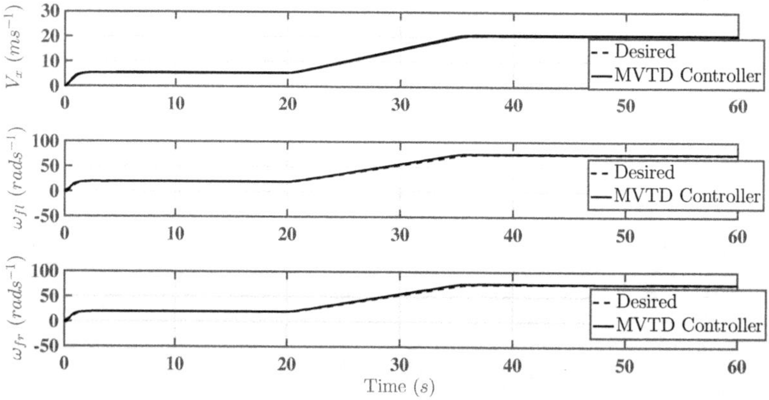

Figure 6.26: Longitudinal states for cornering while acceleration

The Figure 6.27 shows the plot of the vertical tyre loading of the NTTW vehicle during cornering while acceleration. It can be observed that while acceleration (20 to 35 s), the vertical loading of rear wheel increases and simultaneously the loading at the front wheels decreases. This behavior is expected and it can be observed that the control inputs are modified to cope with this change. Figure 6.28 shows the plot of all the body accelerations during this maneuver. The longitudinal acceleration is 0.1 g and the peak perceived lateral acceleration experienced by the driver during this maneuver is 0.023 g. The NTTW vehicle trajectory is plotted in Figure 6.29 where the actual trajectory is able to follow the desired trajectory quite well. Thus, the MVTD controller is able to perform satisfactorily for this case.

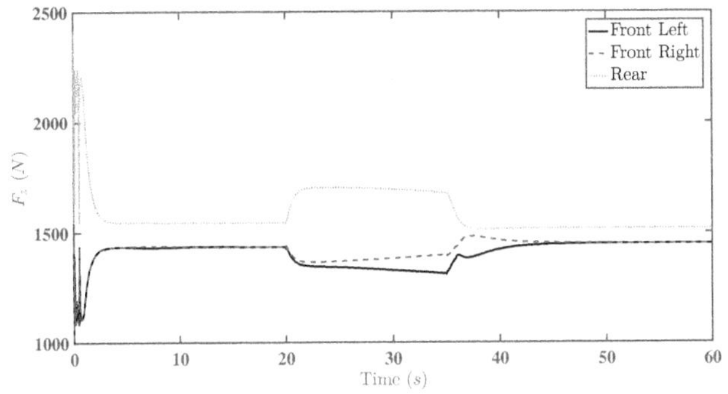

Figure 6.27: Vertical tyre forces for cornering while acceleration

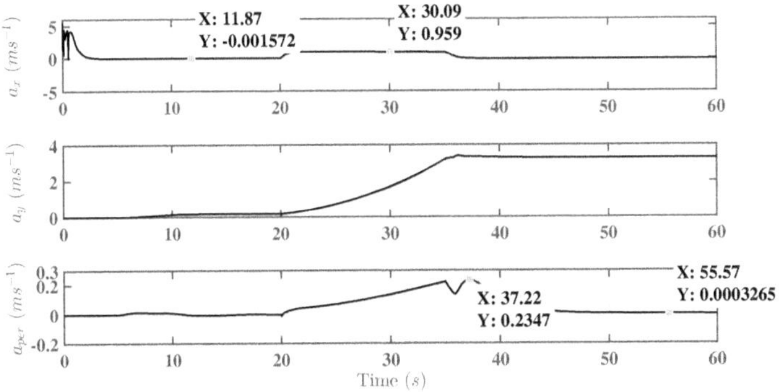

Figure 6.28: Body accelerations for cornering while acceleration

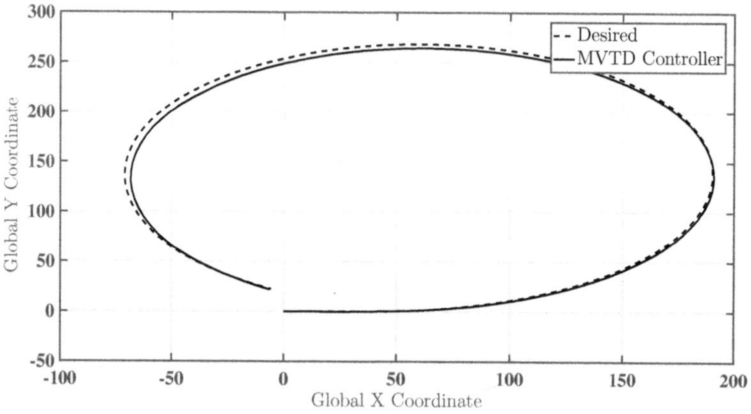

Figure 6.29: Vehicle trajectory for cornering while acceleration

6.4.5 Cornering while deceleration

The braking system is not included in the scope of the thesis, hence a case of minimal deceleration is considered for analysis. The driver (speed command) accelerates from 0 to 30 kmph ($\approx 8ms^{-1}$) and decelerates gradually to 25 kmph ($\approx 7ms^{-1}$) while taking a constant radius of turn of 135 m. Figure 6.30 through Figure 6.34 show plots for the control inputs, the vehicle states, body accelerations and vehicle trajectory. It can be observed from Figure 6.30 that during deceleration (10 to 50 s), the required control inputs increase and then settle when the constant steady state is achieved. Figure 6.31 represents the plot for all the lateral states. As expected, the vehicle tilting angle decreases as the speed of the vehicle decrease from 0.04 rad (2.5 deg) at 30 kmph to 0.037 rad (2.1 deg) at 25 kmph. Further, some peaks are observed during the start of the deceleration period in the states. It is interesting to note that the drive torques also vary during deceleration. The trajectory plot shown in Figure 6.34 depicts the desired versus actual trajectory of the NTTW vehicle during cornering while deceleration.

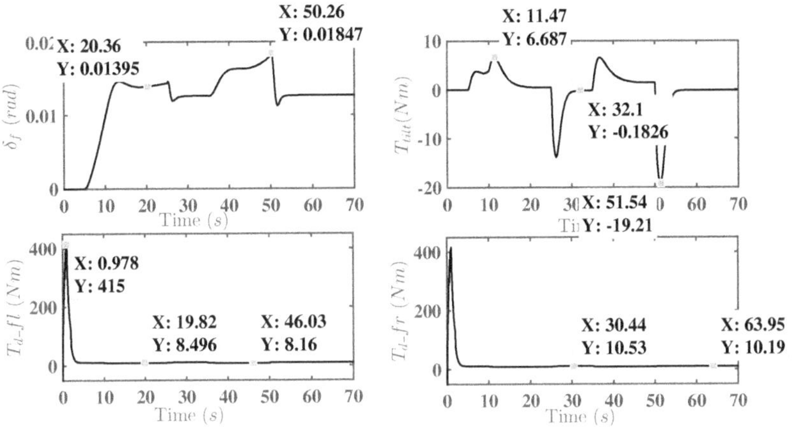

Figure 6.30: Control inputs for cornering while deceleration

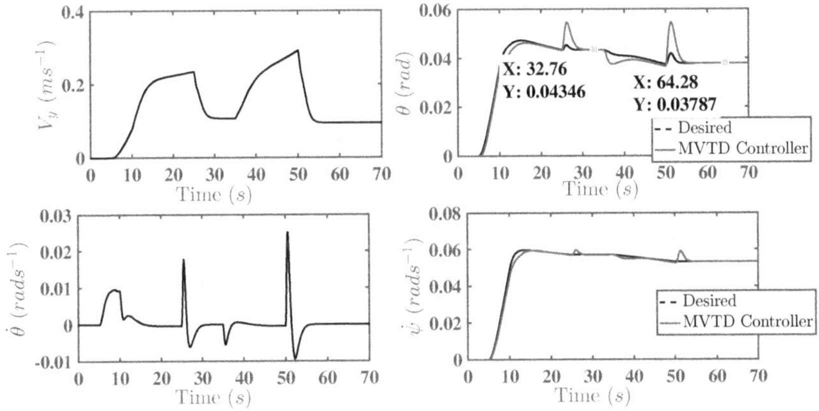

Figure 6.31: Lateral states for cornering while deceleration

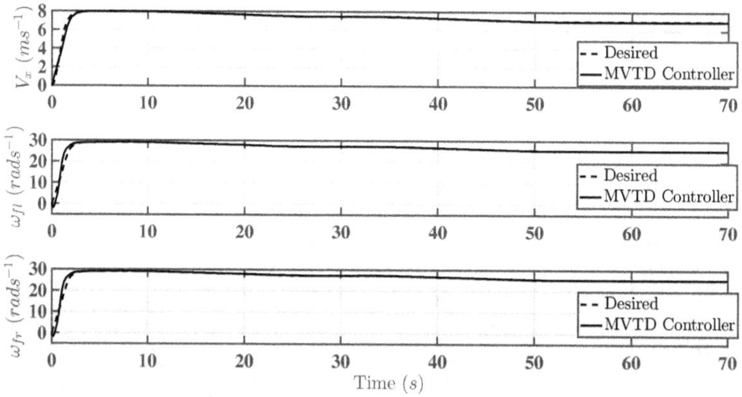

Figure 6.32: Longitudinal states for cornering while deceleration

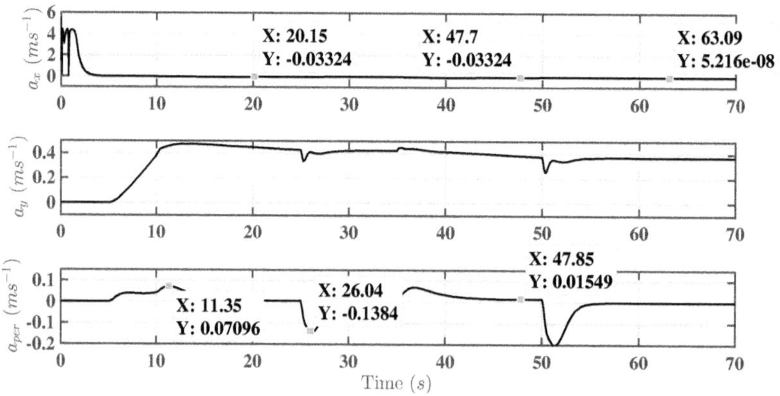

Figure 6.33: Body accelerations for cornering while deceleration

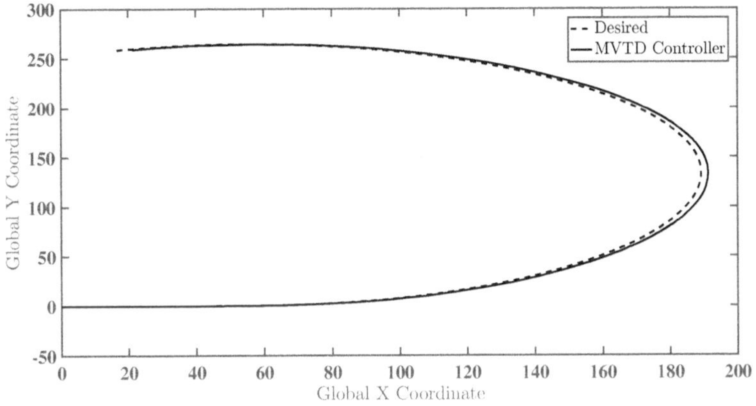

Figure 6.34: Vehicle trajectory for cornering while deceleration

6.4.6 MVTD with estimator

To complete the discussion on the performance of MVTD controller, some plots are presented in this section for MVTD controller with estimator in loop. The estimator is designed using extended Kalman filter (EKF) to reconstruct unmeasured signals of lateral-tilt states of NTTW vehicle. The design of EKF along with all the details regarding its implementation is presented in detail in Chapter 7. A constant radius turn maneuver is simulated in this section. Initially, the vehicle moves in a straight line for 5 sec. Then, the driver takes a left turn of 35 m radius of curvature at 30 kmph ($\approx 8ms^{-1}$) and continues with this constant radius until the end of the simulation. Figure 6.35 through Figure 6.38 show plots for the control inputs, the vehicle lateral and longitudinal states and the estimated states of the lateral-tilt dynamics of the NTTW vehicle. It can be observed from Figure 6.37 that all the states of the lateral-tilt dynamics are reconstructed properly using the estimator.

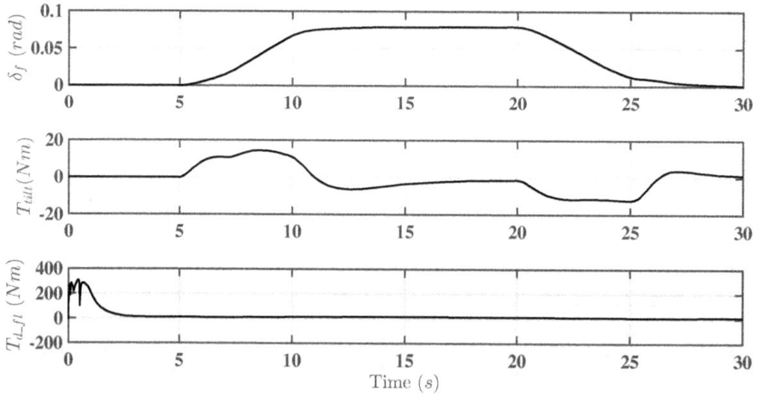

Figure 6.35: Inputs of MVTD controller with estimator

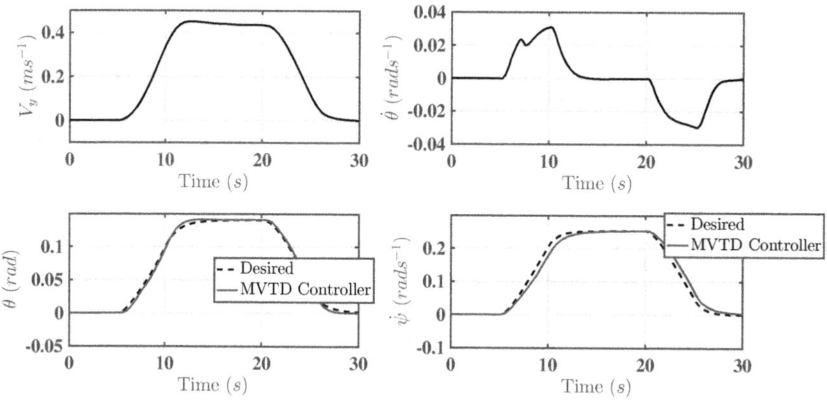

Figure 6.36: Lateral states of NTTW vehicle of MVTD controller with estimator

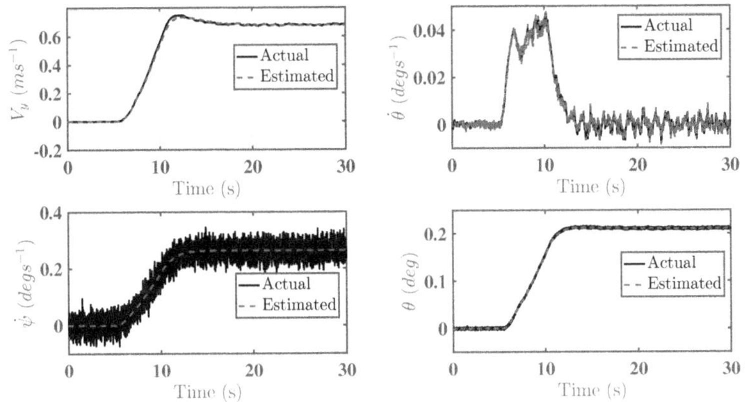

Figure 6.37: Estimation results of MVTD controller using EKF with sensor noise

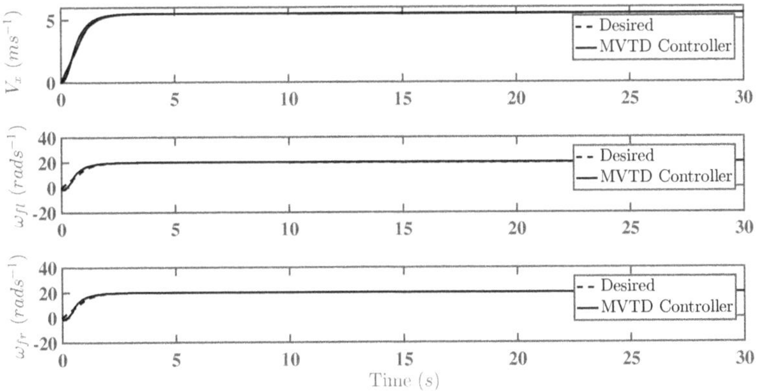

Figure 6.38: Longitudinal states of NTTW vehicle of MVTD controller with estimator

6.5 Summary

i. This chapter presents the development of an integrated multivariable controller based on optimal control strategy (LQR) for the NTTW vehicle.

ii. In this chapter, a multivariable control design is presented for simultaneous control of the lateral-tilt states $(V_y, \dot{\theta}, \theta, \dot{\psi})$ with two inputs (δ_f and T_{tilt}). Similarly, the two drive torques of T_{d_fl} and T_{d_fr} are used to control all the longitudinal states (V_x, ω_{fl} and ω_{fr}) simultaneously. A state feedback control approach is taken to achieve these objectives. Trial and error method is used to tune the individual controllers for lateral-tilt and longitudinal dynamics.

iii. All the states are assumed to be available for feedback while designing this controller. However, from the practical point of view, all the states are not measurable and hence, a suitable estimator design needs to be formulated to accomplish a realistic controller design. This will be discussed in the next chapter.

iv. It is found through simulation studies that the MVTD controller proves to be better than baseline controller. Further simulation studies are carried out using various test maneuvers to analyze the performance of MVTD controller. The MVTD controller provides satisfactory performance in terms of vehicle stability and it is able to achieve stipulated control objectives. However, it is observed that the trajectory tracking needs improvement and the transient performance can be better.

v. Also, it is observed that the drive torques obtained during simulations match well with analytical calculations given in Appendix A.

Chapter 7

Estimation of Unmeasured States for Nonlinear NTTW Vehicle Model

7.1 Introduction

In the previous Chapters 5 and 6, it is assumed that all the states are measured by respective sensors and are available for feedback. However, in practical scenario obtaining full information on all the states is not feasible and only part of the state variable may be accessible. Thus, a straightforward application of full state feedback control is limited. Also, the increased number of sensors makes the overall system more complex in implementation. Thus, there is a need to design a suitable estimator to generate the unmeasured states. State-variable estimation is frequently done in vehicular control systems. The unknown parameters are obtained from a process model rather than an additional sensor [61]. State-variable estimation allows for substantial cost saving by replacing measuring devices.

In this work, it is assumed that the lateral-tilt states of lateral-tilt dynamics ($\dot{\theta}$, θ and $\dot{\psi}$) are measured using sensors such as solid state rate gyros and inclinometers. However, the lateral velocity (V_y) measurements are not acquirable and the lateral acceleration measurements contain a significant amount of noise to integrate. Hence, it is beneficial to design an estimator to generate this signal. The 'Kalman filter' (KF) is an estimation method used to generate non-measurable states and to filter noisy signals. The NTTW vehicle is a nonlinear system and hence, an extended Kalman filter (EKF), which is an advancement of KF is used in this research work. The EKF estimates the unmeasured state signals using the non-linear system model and the measured state signals. An EKF is designed for the lateral-tilt model to reconstruct the four states of the system (V_y, $\dot{\theta}$, θ and $\dot{\psi}$). In the context of NTTW, authors in [51] have presented a design of a static calculator to estimate the unmeasured state using available measurements. However, estimation in the presence of noisy measurements is not discussed, which is an unavoidable practical occurrence. In this chapter, EKF design is presented and its performance in presence of noisy sensor signals is also addressed. Some initial results of Kalman filter are presented to test its performance for the NTTW vehicle. The estimation results of EKF are subsequently presented through simulation studies.

7.2 EKF Algorithm

This section presents the steps involved to design a suitable estimator using EKF technique. The overall block diagram of the implementation is shown in Figure 7.1.

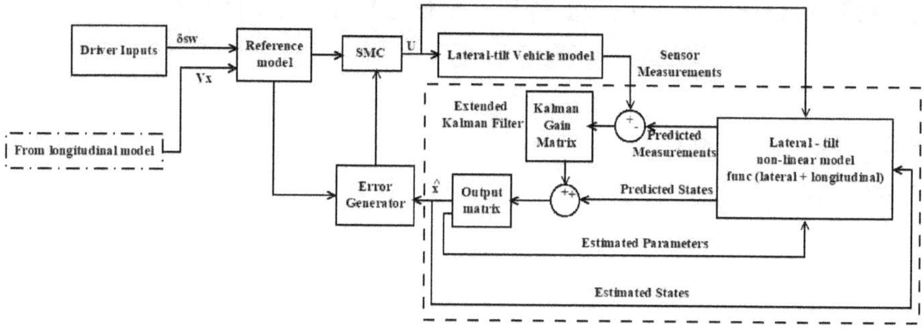

Figure 7.1: Block diagram of estimator design

The EKF is essentially a set of mathematical equations that comprises of two stages, a predictor, and a corrector. The predictor stage predicts future state estimates based on the system dynamics and the corrector stage, corrects the estimates based on some available current measurements [70], [71]. The state-space representation of the non-linear system can be given by (7.1).

State and output equations

$$\dot{\mathbf{x}} = f(\mathbf{x}, \mathbf{u}) + w \tag{7.1}$$

$$\mathbf{y} = h(\mathbf{x}) + v$$

where $f(.)$ and $h(.)$ represent the state and the output functions respectively and are calculated according to system dynamics.

Prediction stage

The first stage of EKF is given by equations (7.2) and (7.3). This consists of predicting the estimate of the current state by projecting the estimate of state and error covariance from previous time period forward in time. For implementation, the proposed estimator is discretized using first order Euler formula. The predicted state estimate is derived by (7.2):

$$\hat{\mathbf{x}}_{k|k-1} = f(\hat{\mathbf{x}}_{k-1}, \mathbf{u}_k) \tag{7.2}$$

where, $f(\hat{\mathbf{x}}_k, \mathbf{u}_k)$ denotes the NTTW nonlinear function of lateral-tilt dynamics which are generated using the estimated states ($\hat{\mathbf{x}}_k$) and the control input ($\mathbf{u}_k$). This non-linear model is a function of lateral-tilt as well as some longitudinal model parameters such as V_x and ω_f due to the coupling of lateral-tilt and longitudinal dynamics. The subscript 'k' denotes the discretized values at specific sample time. The predicted covariance estimate is derived as:

$$P_{k|k-1} = F_k P_{k-1} F_k^T + \mathbb{Q}_k \tag{7.3}$$

where F_k denotes the linearized NTTW lateral-tilt vehicle state matrix in discrete domain, which is calculated online during the complete control iteration cycle. The non-linear mathematical model of lateral-tilt dynamics (presented in Section 3.6.1) is linearized about $\theta = 0$ to obtain F matrix [47].

$$F = \begin{bmatrix} \frac{1}{V_x}(-\frac{p}{m_{cg}} - \frac{p{h_{cg}}^2}{I_{vehicle\_xx}}) & 0 & t(\frac{1}{m_{cg}} + \frac{{h_{cg}}^2}{I_{vehicle\_xx}}) - \frac{m_{cg} g {h_{cg}}^2}{I_{vehicle\_xx}} & \frac{1}{V_x}(-\frac{q}{m_{cg}} - \frac{q{h_{cg}}^2}{I_{vehicle\_xx}}) - V_x \\ \frac{1}{V_x}\frac{p h_{cg}}{I_{vehicle\_xx}} & 0 & \frac{(m_{cg} g h_{cg}) - (t h_{cg})}{I_{vehicle\_xx}} & \frac{1}{V_x}\frac{q h_{cg}}{I_{vehicle\_xx}} \\ 0 & 1 & 0 & 0 \\ \frac{-1}{V_x}\frac{q}{I_{vehicle\_zz}} & 0 & \frac{s}{I_{vehicle\_zz}} & \frac{-1}{V_x}\frac{r}{I_{vehicle\_zz}} \end{bmatrix}$$

where,

$$p = C_{\alpha\_fl} + C_{\alpha\_fr} + C_{\alpha\_r}$$

$$q = L_f(C_{\alpha\_fl} + C_{\alpha\_fr}) - L_r C_{\alpha\_r}$$

$$r = {L_f}^2(C_{\alpha\_fl} + C_{\alpha\_fr}) + {L_r}^2 C_{\alpha\_r}$$

$$s = L_f(C_{\lambda\_fl} + C_{\lambda\_fr}) - L_r C_{\lambda\_r}$$

$$t = C_{\lambda\_fl} + C_{\lambda\_fr} + C_{\lambda\_r}$$

Corrector stage

The second stage in the EKF algorithm is generation of an updated estimate of state by correcting the predicted estimate of state and by taking the recent state measurement into account. These steps are represented by equations (7.4) through (7.6). The Kalman gain ($\mathbb{K}_k$) is represented as:

$$\mathbb{K}_k = P_{k|k-1} H_k^T (H_k P_{k|k-1} H_k^T + \mathbb{R})^{-1} \tag{7.4}$$

where, H_k denotes the Jacobian of the measurement matrix h at step k. The estimate is updated via current measurements as:

$$\hat{\mathbf{x}}_k = \hat{\mathbf{x}}_{k|k-1} + \mathbb{K}_k(\tilde{\mathbf{y}}_k - h(\hat{\mathbf{x}}_{k|k-1})) \tag{7.5}$$

The predicted covariance estimate is then updated to:

$$P_k = (I - \mathbb{K}_k H_k) P_{k|k-1} \tag{7.6}$$

The current state estimate, $\hat{\mathbf{x}}_{k+1}$ is obtained after both the prediction and correction stages are implemented. The output state estimate, $\mathbf{y}_k$ (given by equation (7.7)), can be directly calculated through $\hat{\mathbf{x}}_k$.

$$\mathbf{y}_k = H_k \hat{\mathbf{x}}_k \tag{7.7}$$

Then, the values of $\hat{\mathbf{x}}_k$ and P_k are recursively used in the next predictor stage.

In this work,

- k denotes a discrete point in time
- $\mathbf{u}_k$ is input vector, viz. (δ_f and T_{tilt})
- $\mathbf{x}_k$ is a vector of the actual states, which may be observable but not measured ($V_y, \dot{\theta}, \theta, \dot{\psi}$)
- $\hat{\mathbf{x}}_k$ is a vector of estimated states
- $\mathbf{y}_k$ is a vector of the actual process outputs
- $\tilde{\mathbf{y}}_k$ is a vector of the measured states ($\dot{\theta}$, θ, and $\dot{\psi}$)
- w and v are process and observation noises respectively. They are assumed to be zero mean Gaussian with covariance $\mathbb{Q}_k$ and $\mathbb{R}_k$ respectively. The values of $\mathbb{Q}_k$ and $\mathbb{R}_k$ are obtained by tuning and are given in Table 7.1
- P_k is an estimate of the covariance of the measurement error and $\mathbb{K}_k$ is the Kalman gain.

Table 7.1: Values of error covariance matrix for EKF

Covariance	States			
	V_y	$\dot{\theta}$	θ	$\dot{\psi}$
$\mathbb{Q}_k$	1×10^{-5}	1×10^{-4}	1×10^{-4}	1×10^{-5}
$\mathbb{R}_k$	-	0.01	0.01	0.05

7.3 Results and discussion

In this section, estimation results of Kalman filter (KF) are presented to test its efficacy for generating state estimates for lateral-tilt dynamics of the NTTW vehicle. Further, simulation results of EKF are presented in detail.

7.3.1 Kalman filter

A constant radius turn maneuver with around 75 m of turning radius and constant speed of 30 kmph is simulated to test the functionality of KF and is shown in Figure 7.2. The observer design based on KF uses linearized system dynamics to calculate the unmeasured signals from the measured signals. The values of $\mathbb{Q}'_k$ and $\mathbb{R}'_k$ for KF are obtained by tuning and are given in Table 7.2. It is observed that although all the measured signals are generated properly, the unmeasured estimated state V_y has a considerable amount of offset from the actual signal. Similarly, a double lane change maneuver with around 23 m of turning radius and constant speed of 20 kmph ($\approx 5.5ms^{-1}$) is simulated to further test the operation of KF and is shown in Figure 7.3. The performance of the KF further deteriorates for this test case as the reconstructed signals of V_y, $\dot{\theta}$, and $\dot{\psi}$ have a constant error between actual and estimated signals. This is due to the fact that there is nonlinearity present in lateral-tilt dynamics.

Table 7.2: Values of error covariance matrix for KF

Covariance	States			
	V_y	$\dot{\theta}$	θ	$\dot{\psi}$
$\mathbb{Q}'_k$	1×10^{-3}	0.0016	0.00025	1×10^{-2}
$\mathbb{R}'_k$	-	1×10^{-4}	1×10^{-4}	1×10^{-4}

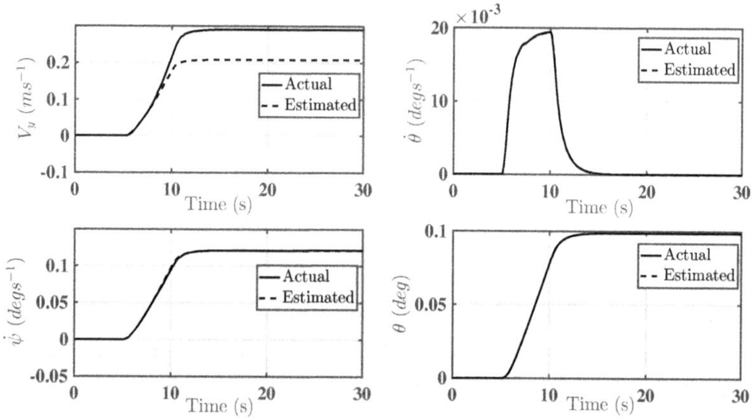

Figure 7.2: Estimation results using KF for constant radius turn

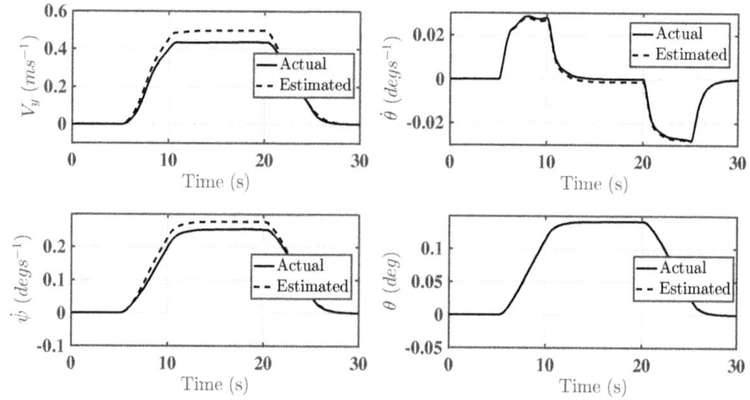

Figure 7.3: Estimation results using KF for a double lane change

Hence, it implies that the standard KF proves to be inadequate to estimate the states of nonlinear dynamics of the NTTW vehicle. The same maneuvers have been simulated with EKF to analyze its performance.

7.3.2 Extended Kalman filter

To test the operation of EKF, a constant radius turn maneuver with around 75 m of turning radius and constant speed of 30 kmph is simulated. Initially, the vehicle moves in a straight line for 5 sec. Then, the driver takes a left turn of 75 m radius of curvature at 30 kmph and continues at the constant radius until the end of the simulation. The EKF is able to generate a reasonable estimate of the unmeasured state (V_y) using the three measured states. In Figure 7.4, the actual and estimated states are compared. The actual states depict the ideal signals that are expected from the NTTW vehicle dynamics. It can be observed from Figure 7.4 that all the four states are estimated correctly without a bias or drift.

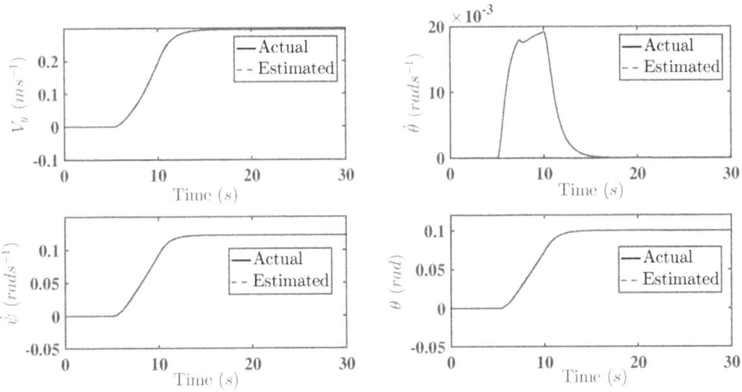

Figure 7.4: Estimation results using EKF without noise

Further, in Figure 7.5 EKF outputs are shown when white gaussian noise is added to actual states to emulate the presence of sensor noise. To test the operation of EKF, a double lane change maneuver with around 23 m of turning radius and constant speed of 20 kmph ($\approx 5.5ms^{-1}$) is simulated. Initially, the vehicle moves in a straight line for 5 sec. Then, the driver takes a turn of 23 m radius of curvature, the transient is between 5 to 10 seconds. From 10 to 20 seconds, the driver takes a constant radius of turn. At 20 seconds, the driver changes his direction of travel to a straight line till the end of the simulation. Table 7.3 shows the noise standard deviations for each sensor which are considered in this work. The plot shown in Figure 7.5 represents the comparison of measured noisy signals and the filtered estimated signals. It can be observed that after some tuning of gains $\mathbb{Q}_k$ and $\mathbb{R}_k$, the EKF is able to reject the sensor noise satisfactorily.

Table 7.3: Standard deviation of noise

Parameter	Noise Deviation	Unit
$\dot{\theta}$	0.001	$rads^{-1}$
θ	0.01	rad
$\dot{\psi}$	0.2236	$rads^{-1}$

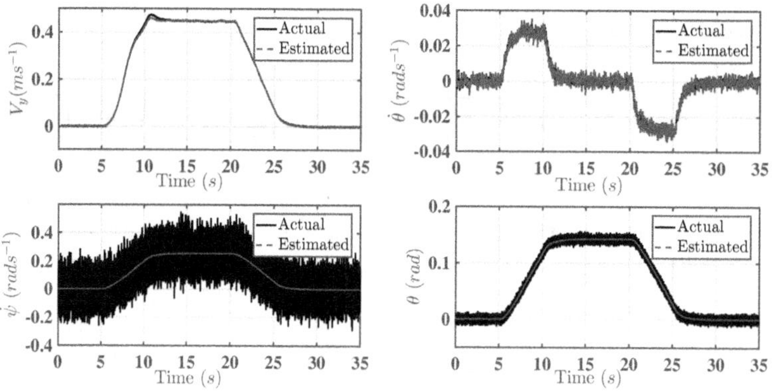

Figure 7.5: Estimation results using EKF with noise

It can be observed from Figure 7.4 and Figure 7.5 that the measured and estimated states match to a great extent in all the states. However, comparatively the noise cancelation in $\dot{\theta}$ is less than the others as it is a derivative state which tends to amplify noise. Nevertheless, it can be observed in later chapters that the effect of this noise is minimal on the response of the NTTW vehicle stability controller. This estimator thus can be used in a feedback loop to provide the state signals to the controller. The Figure 7.6 and Figure 7.7 show the plots of Kalman gain matrix ($\mathbb{K}_k : 4 \times 3$) and process covariance matrix ($\mathbb{P}_k : 4 \times 4$) respectively. It can be observed from the plots that the gain values settle down in time (steady state values are marked on the plots) indicating that the EKF is stable.

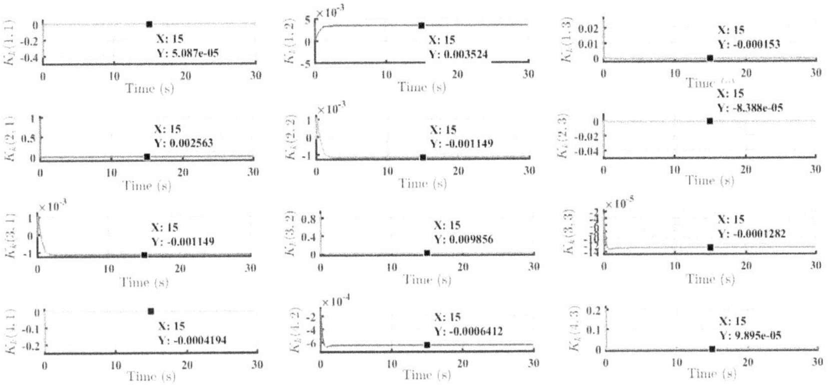

Figure 7.6: Kalman gain matrix

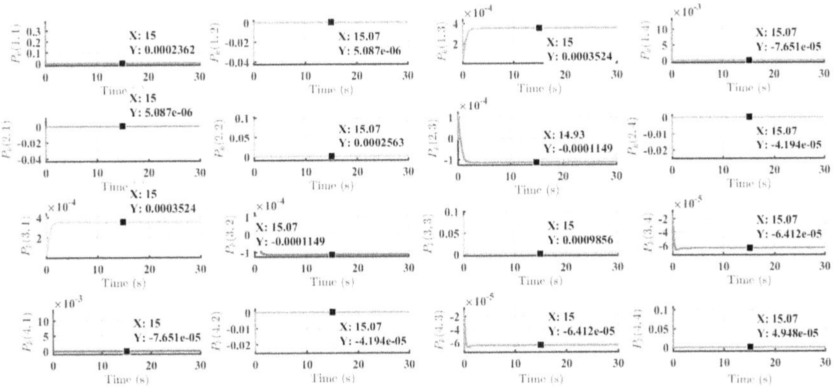

Figure 7.7: Covariance gain matrix

Furthermore, the performance of EKF is tested in case of parameter variations. Figure 7.8 and Figure 7.9 show the plots of the responses of the EKF with 20% variation in h_{cg} and L_{tw} respectively. It can be observed that the EKF is able to perform satisfactorily with parameter variation and hence is robust.

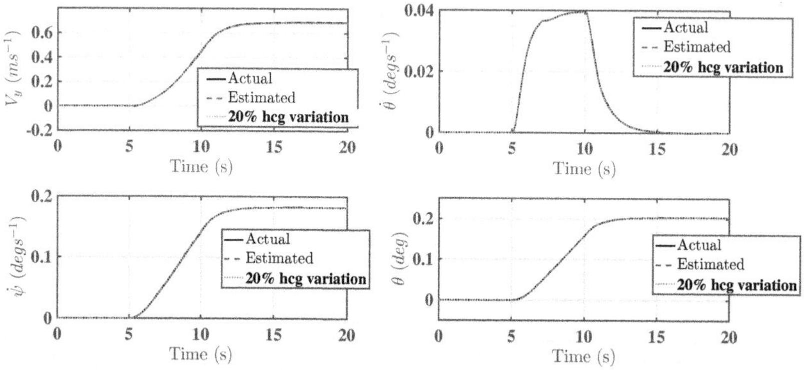

Figure 7.8: Estimation result with variation in h_{cg}

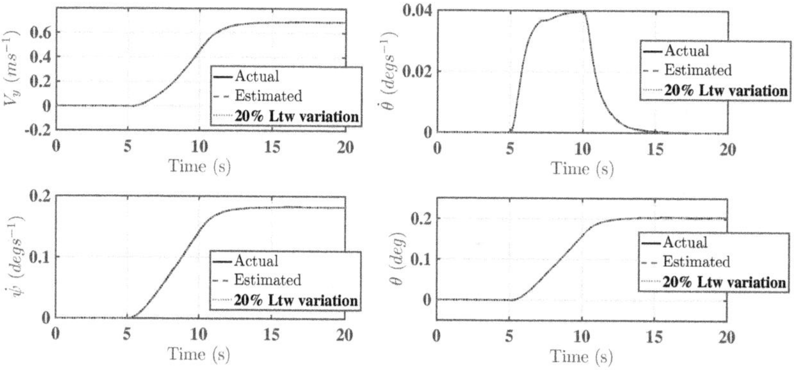

Figure 7.9: Estimation result with variation in L_{tw}

7.4 Summary

i. Initially an estimator is designed using Kalman filter technique to reconstruct the state signals. It is observed from the results that the standard KF is inadequate to successfully estimate the states of nonlinear dynamics of the NTTW vehicle.

ii. A recursive online estimation technique is designed using EKF methodology for nonlinear lateral-tilt dynamics of the NTTW vehicle. It is observed that the EKF is able to provide noise-free estimates of all the lateral-tilt states of the NTTW vehicle satisfactorily. In addition, EKF is found to be robust to parameter uncertainties.

iii. The EKF is subsequently used to estimate the unmeasured states along with the controllers proposed in this work.

Chapter 8

Design of Integrated Multivariable Sliding Mode (IM-SM) Controller Strategy

8.1 Introduction

The MVTD controller, which is based on optimal control strategy, is not robust to unmodeled dynamics. It is also observed that the transient performance of the MVTD controller needs improvement which will further improve the trajectory tracking performance as well. Due to highly nonlinear nature of the plant under consideration, there was a necessity to design a suitable controller which is robust to modeling uncertainties. Sliding mode control (SMC) is known to be robust to unmodeled dynamics and parametric uncertainties and hence to further enhance the performance of stability control of NTTW platform, SMC is investigated. The authors in [45] and [72] have designed tilt angle and speed controllers using sliding mode framework for a narrow tilting three wheeled vehicle. Furthermore, the authors have proposed a sliding mode control for tracking of desired steering angle and velocity of the NTTW vehicle. The NTTW vehicle is modeled as nine DoF nonlinear using multibody approach. They have used a nonlinear model for control design. On the contrary, the work in this thesis has used a linearized version of the lateral-tilt and longitudinal model individually for the proposed control design. This approach is more useful for implemention of control algorithm on an embedded platform. Additionally, the multivariable controller designed is this thesis provides a better control over the states of the NTTW vehicle, such as yaw rate and rate of change of tilt angle, which may provide a better tracking of the desired trajectory. In this chapter, an integrated multivariable sliding mode (IM-SM) controller is proposed for the stability of

NTTW vehicle. Similar to MVTD controller configuration, IM-SM controller consists of two multivariable controllers, one for lateral-tilt dynamics and the other for longitudinal dynamics. The proposed controller design allows simultaneous control of the lateral-tilt and longitudinal dynamic states in an integrated manner to achieve the desired performance.

Initially, background on sliding mode control strategy is presented in brief and then the design of proposed IM-SM controller strategy is explained in detail. Additionally, stability analysis of the designed sliding surface of lateral-tilt and longitudinal dynamics are reported to ascertain finite time convergence. Then, extensive simulation studies are performed using chosen test maneuvers for the performance analysis of the NTTW vehicle.

8.2 Design and implementation of IM-SM controller

In this work, SMC based approach is proposed for dynamic stability control of NTTW vehicle. The overall architecture of the proposed integrated multivariable sliding mode (IM-SM) controller is represented in Figure 8.1. Two different multivariable controllers based on sliding mode technique for non-linear lateral-tilt and longitudinal dynamics are integrated to achieve the desired performance. The stability of proposed IM-SM controller is evaluated using Lyapunov analysis.

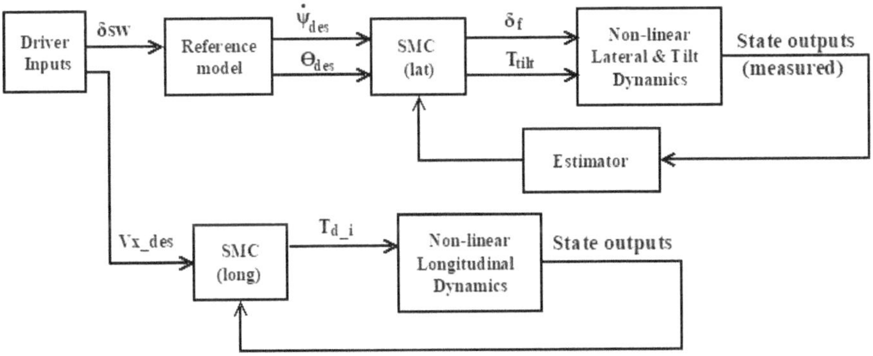

Figure 8.1: The architecture of the IM-SM controller

8.2.1 Design of multivariable SMC for lateral-tilt dynamics

The sliding mode design approach consists of two stages:

i. The design of a sliding surface/manifold such that it is stable

ii. The design of control law which will ensure finite time reachability of system dynamics to the sliding surface

The design of the sliding manifold can be linear or nonlinear depending on the specified goal. The simplest manifold is a linear one which is considered in this work. The design of a sliding surface plays a key role in formulating a sliding mode controller. A particular canonical form, also known as the regular form is considered to design a stable sliding manifold. Consider the nominal linear model for lateral-tilt dynamics given as:

$$\dot{\mathbf{x}}_{lat} = A_{lat}\mathbf{x}_{lat} + B_{lat}\mathbf{u}_{lat} \tag{8.1}$$

Where B_{lat} has full rank and (A_{lat},B_{lat}) is a controllable pair. Then, the sliding surface for lateral-tilt dynamics is designed as:

$$S_{lat} = M_{lat}(\mathbf{x}_{lat} - \mathbf{x}_{lat\_des}) \tag{8.2}$$

Here, $M_{lat} \in \Re^{m \times n}$ and m and n represent the number of inputs and outputs (here, $m = 2$ and $n = 6$ for lateral-tilt dynamics) respectively. A modification is proposed in the surface design to ascertain zero steady-state error while tracking of the desired maneuver. The system matrix for the lateral-tilt dynamics controller is modified ($[A_{lat}, B_{lat}, \mathbf{x}_{lat} \Longrightarrow A'_{lat}, B'_{lat}, \mathbf{x}'_{lat}]$) by augmenting state vector with two integral error states (θ and $\dot{\psi}$). Hence,

$$\mathbf{x}'_{lat} - \mathbf{x}'_{lat\_des} = \begin{bmatrix} V_y & e_{\dot{\theta}} & e_{\theta} & e_{\dot{\psi}} & \int e_{\theta} & \int e_{\dot{\psi}} \end{bmatrix}^T$$

Defining,

$$e_{lat} = \mathbf{x}'_{lat} - \mathbf{x}'_{lat\_des}$$

where, e_{lat} are the lateral-tilt dynamics state error signals which are obtained by subtracting actual states (V_y,$\dot{\theta}$,θ and $\dot{\psi}$) from desired states (as obtained in Section 4.3). The desired values of $\dot{\theta}$ and integral states are zero. The pattern of V_y state is expected to follow the input to the steering wheel angle of the vehicle. The tracking of V_y is not a part of the control objectives of this work and hence is not addressed further. The system is transformed using transformation matrix T_r which is obtained by performing QR decomposition on the input matrix. The system is hence transformed as:

$$z = T_r e_{lat}$$

Such that:

$$T_r B'_{lat} = [0 \;\; B_2]^T$$

where, $B_2 \in \Re^{m \times m}$ is input matrix. After transformation, the configuration (8.1), (8.2) may be expressed in the regular form as:

$$\dot{\mathbf{z}_1} = A_{11}\mathbf{z}_1 + A_{12}\mathbf{z}_2 \tag{8.3}$$

$$\dot{\mathbf{z}_2} = A_{21}\mathbf{z}_1 + A_{22}\mathbf{z}_2 + B_2\mathbf{u}_{lat} \tag{8.4}$$

where $[z_1 \ \ z_2]^T$ is the transformed state vector and $z_1 \in \Re^{n-m \times 1}$ and $z_2 \in \Re^{m \times 1}$. A_{11}, A_{12}, A_{21} and A_{22} are the reduced order transformed system matrices. The sliding manifold is transformed as:

$$S_{lat} = M_{lat_1}\mathbf{z}_1 + M_{lat_2}\mathbf{z}_2 \tag{8.5}$$

This transformation is necessary to design M_{lat_1} and M_{lat_2} for stable consideration. The lateral-tilt and longitudinal dynamics are converted into their respective regular forms using the methodology described in earlier steps. "Pole placement method" and "Quadratic minimization" are the two methods to design stable sliding surface/manifold [73]. In this work, two separate controllers are designed for lateral-tilt and longitudinal dynamics. Hence, the design of sliding manifold and the formulation of control law are also performed individually for each system.

Design of sliding surface matrix for lateral-tilt dynamics (M_{lat})

The parameter M_{lat} is designed such that S_{lat} is stable. The design matrix M_{lat} for lateral-tilt dynamics is obtained using quadratic minimization method [73], [74]. According to quadratic minimization method, a linear quadratic cost function is minimized to obtain optimal surface constants. Consider the problem of minimizing the quadratic performance index (J_{lat}):

$$J_{lat} = \frac{1}{2}\int_{t_s}^{\infty} \mathbf{x}_{lat}^T Q_{lat} \mathbf{x}_{lat} dt \tag{8.6}$$

where Q_{lat} is both symmetric and positive definite and t_s is the time at which the sliding motion commences. The aim is to minimize equation (8.6). Transforming weighting matrix to regular form coordinate system by:

$$Q_t = T_r Q_{lat} {T_r}^T \tag{8.7}$$

where,

$$Q_t = \begin{bmatrix} Q_{11} & Q_{12} \\ Q_{21} & Q_{22} \end{bmatrix} \tag{8.8}$$

The associated weighting matrices are defined as follows:

$$\hat{Q}_{lat} = Q_{11} - Q_{12}{Q_{22}}^{-1}Q_{21} \tag{8.9}$$

$$\hat{A}_{lat} = A_{11} - A_{12}{Q_{22}}^{-1}Q_{21} \tag{8.10}$$

The controllability of the original (A_{lat},B_{lat}) pair ensures that the pair (A_{11},A_{12}) is controllable [73] and this, in turn, is sufficient to ensure the controllability of ($\hat{A}_{lat}$,A_{11}). The LQR problem is solved using:

$$[K, P_1, E] = lqr(\hat{A}_{lat}, A_{12}, \hat{Q}_{lat}, Q_{22})$$

Then, the sliding surface matrix in terms of the original coordinates is obtained by [73] as follows:

$$G_{lat} = Q_{22}^{-1}(A_{12}{}^T P_1 + Q_{21}) \tag{8.11}$$

$$M_{lat} = I_{m\times m}[G_{lat}I_{m\times m}]T_r \tag{8.12}$$

where $I_{m\times m}$ is identity matrix. Accordingly, the surface matrix for lateral-tilt dynamics is designed as:

$$M_{lat} = \begin{bmatrix} 0.07 & 1.12 & 10.1 & -0.98 & 20.86 & -1.67 \\ 0.05 & -0.42 & -3.77 & -1.04 & -7.74 & -12.57 \end{bmatrix}$$

Synthesis of control law for lateral-tilt dynamics

The next stage in the design of the IM-SM controller is to formulate the control law using Gao's reaching law [73] which guarantees finite time reaching to the surface. The Gao's proportional and constant rate law is:

$$\dot{S}_{lat} = -k_{lat}sgn(S_{lat}) - qS_{lat} \tag{8.13}$$

Where k_{lat} and q are diagonal matrices of order $m \times m$. The values of the positive constants (k_{lat} and q) are suitably selected as will be discussed in Section 8.4. The signum function is denoted by $sgn(.)$. The equivalent control law can be derived by substituting (8.13) in the derivative of (8.2) as:

$$\begin{aligned} -k_{lat}sgn(S_{lat}) - qS_{lat} &= M_{lat}(\dot{\mathbf{x}}'_{lat} - \dot{\mathbf{x}}'_{lat\_des}) \\ -k_{lat}sgn(S_{lat}) - qS_{lat} &= M_{lat}(A'_{lat}\mathbf{x}'_{lat} + B'_{lat}\mathbf{u}_{lat}) - M_{lat}(\dot{\mathbf{x}}'_{lat\_des}) \end{aligned} \tag{8.14}$$

Accordingly after solving (8.14):

$$\mathbf{u}_{lat} = (M_{lat}B'_{lat})^{-1}[-M_{lat}A'_{lat}\mathbf{x}'_{lat} - k_{lat}sgn(S_{lat}) - qS_{lat} + M_{lat}\dot{\mathbf{x}}'_{lat\_des}] \tag{8.15}$$

Hence, (8.15) represents the control law of lateral-tilt dynamics.

8.2.2 Design of multivariable SMC for longitudinal dynamics

Similar to the previous section, stable sliding surface and control law are designed for longitudinal dynamics.

Design of sliding surface matrix for longitudinal dynamics (M_{long})

The sliding surface matrix (M_{long}) design for longitudinal dynamics is performed using quadratic minimization method. M_{long} is designed such that sliding surface for longitudinal dynamics (S_{long}) is stable. The sliding surface S_{long} is designed as:

$$S_{long} = M_{long}(\mathbf{x}_{long} - \mathbf{x}_{long\_des}) \tag{8.16}$$

where $M_{long} = \begin{bmatrix} 1 & 0 & 50.52 \\ 0 & 1 & 50.52 \end{bmatrix}$ and

$$\mathbf{x}_{long} - \mathbf{x}_{long\_des} = \begin{bmatrix} e_{V_x} & e_{\omega_{fl}} & e_{\omega_{fr}} \end{bmatrix}^T$$

Defining,

$$e_{long} = (\mathbf{x}_{long} - \mathbf{x}_{lat\_long})$$

where e_{long} are the error signals of longitudinal dynamics which are obtained by subtracting actual states from desired states (as obtained in Section 4.3).

Synthesis of control law for longitudinal dynamics

Similar to lateral-tilt dynamics, Gao's proportional rate reaching law is applied to longitudinal dynamics and is given as:

$$\dot{S}_{long} = -k_{long} sgn(S_{long}) \tag{8.17}$$

where k_{long} is diagonal matrices of order $m' \times m'$ and m' denotes the number of inputs. The equivalent control law can be derived by substituting (8.17) in the derivative of (8.16) by:

$$\begin{aligned} -k_{long} sgn(S_{long}) &= M_{long}(\dot{\mathbf{x}}_{long} - \dot{\mathbf{x}}_{long\_des}) \\ -k_{long} sgn(S_{long}) &= M_{long}(A_{long}\mathbf{x}_{long} + B_{long}\mathbf{u}_{long}) - M_{long}(\dot{\mathbf{x}}_{long\_des}) \end{aligned} \tag{8.18}$$

Accordingly by solving (8.18):

$$\mathbf{u}_{long} = (M_{long}B_{long})^{-1}[-M_{long}A_{long}\mathbf{x}_{long} - k_{long} sgn(S_{long}) + M_{long}\dot{\mathbf{x}}_{long\_des}] \tag{8.19}$$

Hence, equation(8.19) represents the control law of longitudinal dynamics.

8.3 Stability Analysis

The stability analysis of the proposed controller is performed to ensure that it facilitates finite time convergence of surface to zero. The stability is evaluated by employing Lyapunov analysis and the limiting conditions for gain parameters for the controllers are derived suitably based on Lyapunov criteria.

8.3.1 Finite time convergence of control law for lateral-tilt dynamics

Equation (4.6) is rewritten considering the bounded lumped disturbance present in the lateral-tilt dynamics(η) which includes system uncertainties and external disturbances for the stability proof [75]. This representation is not used in the controller design as η is unknown [76]. However, it is assumed that η is bounded.

$$\dot{\mathbf{x}}'_{lat} = A'_{lat}\mathbf{x}'_{lat} + B'_{lat}\mathbf{u}_{lat} + \eta \tag{8.20}$$

Taking the derivative of (8.2) and combining with (8.20) provides

$$\begin{aligned} \dot{S}_{lat}(i) &= M_{lat}(i,1)\dot{V}_y + M_{lat}(i,2)\ddot{\theta} + M_{lat}(i,3)e_{\dot{\theta}} + M_{lat}(i,4)e_{\ddot{\psi}} + M_{lat}(i,5)\int e_{\dot{\theta}}dt \\ &\quad + M_{lat}(i,6)\int e_{\ddot{\psi}}dt + \eta(i) \end{aligned} \tag{8.21}$$

Here, M_{lat}(i,j) denotes the surface matrix and where, 'i = 1 or 2' denotes the 1st and the 2nd row of the surface matrix and 'j' varies from 1 to 6, denoting each column of the surface matrix. Expanding the error terms in (8.21):

$$\begin{aligned} \dot{S}_{lat}(i) &= M_{lat}(i,1)\dot{V}_y + M_{lat}(i,2)\ddot{\theta} + M_{lat}(i,3)\dot{\theta} + M_{lat}(i,4)\ddot{\psi} + M_{lat}(i,5)\theta + M_{lat}(i,6)\dot{\psi} \\ &\quad -(M_{lat}(i,5)\theta_{des} + M_{lat}(i,6)\dot{\psi}_{des}) + \eta(i) \end{aligned} \tag{8.22}$$

$$\dot{S}_{lat}(i) = M_{lat}(i,1)\dot{V}_y + M_{lat}(i,2)\ddot{\theta} + M_{lat}(i,3)\dot{\theta} + M_{lat}(i,4)\ddot{\psi} + M_{lat}(i,5)\theta + M_{lat}(i,6)\dot{\psi} + |\Gamma(i)| + \eta(i) \tag{8.23}$$

where,

$$|\Gamma(i)| = M_{lat}(i,5)\theta_{des} + M_{lat}(i,6)\dot{\psi}_{des}$$

Theorem 8.3.1. *If the SM controller for lateral-tilt dynamics is designed as equation (8.15) with* $k_{lat}(m) > (|\Gamma(i)| + \eta(i))$ *and* $q(m) > 0$ *(for* $\forall$ *i), then* S_{lat} *will converge to the origin in a finite time.*

Proof. Substituting (8.15) in expanded version of (8.23) produces:

$$\dot{S}_{lat}(i) = -k_{lat}(m)sgn(S_{lat}) - q(m)S_{lat} + (\Gamma(i) + \eta(i)) \tag{8.24}$$

Select a Lyapunov function as $V_{lat}(S_{lat}) = \frac{1}{2}S_{lat}^T S_{lat}$. To ensure sliding, the derivative of $V_{lat}(S_{lat})$ must be negative definite, which means:

$$\begin{aligned} S_{lat}^T \dot{S}_{lat} &< 0 \\ \Rightarrow S_{lat}(i)\dot{S}_{lat}(i) &< 0 \quad \forall i \end{aligned}$$

Solving further by taking the derivative of $V_{lat}(S_{lat})$ along system (8.24) yields

$$\begin{aligned}
\dot{V}_{lat} &= -k_{lat}(m)sgn(S_{lat}(i))S_{lat}(i) - q(m)(S_{lat}^2)(i) + (|\Gamma(i)| + \eta(i))S_{lat}(i) \\
&\leq -k_{lat}(m)|S_{lat}(i)| - q(m)(S_{lat}^2)(i) + (|\Gamma(i)| + \eta(i))|S_{lat}(i)| \\
&\leq -(k_{lat}(m) - (\Gamma(i) + \eta(i)))|S_{lat}(i)| \qquad (8.25)
\end{aligned}$$

Noting that if $k_{lat}(m) > (|\Gamma(i)|+\eta(i))$, then $\dot{V}_{lat} \leq -c_{lat}V_{lat}^{\frac{1}{2}}$ where $c_{lat} = \sqrt{2}(k_{lat}(m)-(\Gamma(i)+\eta(i)))$ (for $\forall$ i). According to finite-time Lyapunov theory, S_{lat} will converge to the origin in finite time [75]. Thus, the parameters are selected based on worst-case analysis of the values of $\Gamma(i)$ and $\eta(i)$:

$$k_{lat} = \begin{bmatrix} 50 & 0 \\ 0 & 10 \end{bmatrix} \text{ and } q = \begin{bmatrix} 10 & 0 \\ 0 & 10 \end{bmatrix}$$ □

8.3.2 Finite time convergence of control law for longitudinal dynamics

Similar to the previous section, equation (4.7) is rewritten considering the bounded lumped disturbance present in the longitudinal dynamics (ζ) which includes system uncertainties and external disturbances for the stability proof. This representation is not used in the controller design as η is unknown [76]. However, it is assumed that ζ is bounded.

$$\dot{X}_{long} = A_{long}X_{long} + B_{long}U_{long} + \zeta \qquad (8.26)$$

Taking the derivative of (8.17) and combining with (8.26) provides

$$\dot{S}_{long}(i) = M_{long}(i,1)e_{\dot{V}_x} + M_{long}(i,2)e_{\dot{\omega}_{fl}} + M_{long}(i,3)e_{\dot{\omega}_{fr}} + \zeta(i) \qquad (8.27)$$

Here, $M_{long}(i,j)$ denotes the surface matrix and where, 'i = 1 or 2' denoting the 1st and the 2nd row of the surface matrix and 'j' varies from 1 to 3, denoting each column of the surface matrix. Expanding the error terms in (8.27):

$$\begin{aligned}
\dot{S}_{long}(i) &= M_{long}(i,1)\dot{V}_x + M_{long}(i,2)\dot{\omega}_{fl} + M_{long}(i,3)\dot{\omega}_{fr} - M_{long}(i,1)\dot{V}_{x_des} \\
&\quad + M_{long}(i,2)\dot{\omega}_{fl_des} + M_{long}(i,3)\dot{\omega}_{fr_des} + \zeta(i)) \qquad (8.28)
\end{aligned}$$

Theorem 8.3.2. *If the SM controller for longitudinal is designed as equation (8.19) with $k_{long}(m') > |\zeta(i)|$, then the sliding manifold S_{long} will converge to the origin in a finite time.*

Proof. Substituting (8.26) in in expanded version of (8.28) produces:

$$\dot{S}_{long}(i) = -k_{long}(m')sgn(S_{long}) + \zeta(i) \qquad (8.29)$$

Select a Lyapunov function as $V_{long}(S_{long}) = \frac{1}{2}S_{long}^2$. Taking the derivative of $V_{long}(S_{long})$ along system (8.29) yields

$$\begin{aligned}
\dot{V}_{long} &= -k_{long}(m')sgn(S_{long}(i))S_{long}(i) + \zeta(i)S_{long}(i) \\
&\leq -k_{long}(m')|S_{long}(i)| + |\zeta(i)||S_{long}(i)| \\
&\leq -(k_{long}(m') - \zeta)|S_{long}(i)| \qquad (8.30)
\end{aligned}$$

Noting that $k_{long}(m') > \zeta$, one has $\dot{V}_{long} \leq -c_{long}V_{long}^{\frac{1}{2}}$ with $c_{long} = \sqrt{2}(k_{long}(m') - \zeta)$ (for $\forall i$). According to finite-time Lyapunov theory, if $c_{long} > 0$, then S_{long} will converge to the origin in finite time. Subsequently, the value of the positive parameter (k_{long}) is suitably selected such that the effect of the perturbations is suppressed as:

$$k_{long} = \begin{bmatrix} 500 & 0 \\ 0 & 500 \end{bmatrix} \qquad \square$$

Remark 2. The control laws (8.15) and (8.19) are discontinuous, which may cause chattering problem which is undesirable [77]. This problem can be sorted by substituting the discontinuous switching function (i.e., sgn(S)) with a sigmoid function. The sigmoid function is essentially a saturation function sat(S) with the boundary layer thickness (Δ) performing a continuous approximation of a signum function. Δ is suitably selected to eliminate chattering without affecting the control performance.

Remark 3. The controller requires information of all states and hence, the estimator is used to generate all the states of the lateral-tilt dynamics from the measured signals as explained in Chapter 7.

8.4 Results and discussion

In this section, a detailed simulation analysis for the performance of the IM-SM controller for the selected maneuvers (as discussed in Section 8.3) has been documented. The subsequent subsections present a detailed analysis of the IM-SM controller through extensive simulations. The test maneuvers discussed in Table 4.2 are used in the analysis throughout the thesis work to maintain uniformity. The sampling time of the IM-SM controller is selected 2 ms and the EKF is simulated at 0.2 ms.

8.4.1 Double lane change maneuver

This section presents the simulation results of the IM-SM controller for various test scenarios. The first case of simulation is performed to show the complete set of dynamic responses of the NTTW vehicle during a double lane change maneuver. Initially, the vehicle moves in a straight line for 5 sec. Then, the driver takes a left turn of 10 m radius of curvature at a constant speed of 10 kmph ($\approx 2.8ms^{-1}$). At 22 seconds, the driver changes his direction of travel to a straight line till the end of the simulation. Figure 8.2 to Figure 8.11 show plots for the control inputs, the vehicle states, tyre forces, tyre slip angles, and the body accelerations.

The δ_{sw} is modified to δ_f by 2.1% to attain the driver's desired trajectory while the vehicle is tilting appropriately to achieve stability. The tilting torque transiently increases then settles to around 2 Nm as steady state is achieved. The peak drive motor torques for front and right wheels are 284 Nm and average torque is around 12.4 Nm during cornering. However, as the vehicle travels in a straight path the average torque is 8.2 Nm. One of the important difference that can be observed in the performance of IM-SM controller from the MVTD controller is that the settling time of IM-SM controller is faster (2 s). The settling time of MVTD controller (Figure 6.10) for this maneuver was quite slow (20 s).

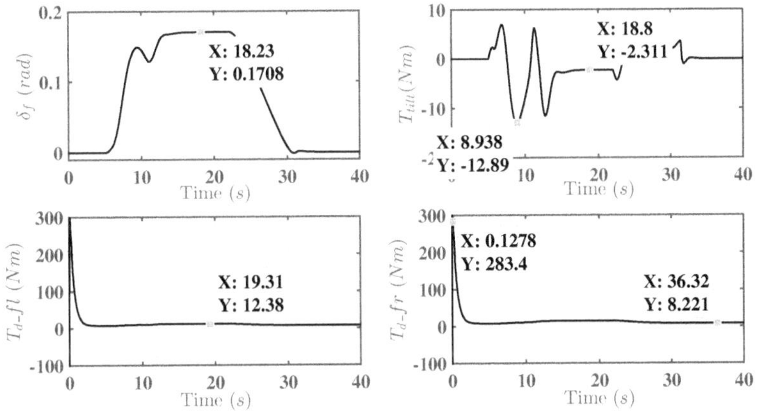

Figure 8.2: Control inputs for double lane change maneuver

All the states (lateral-tilt and longitudinal dynamics) of the NTTW vehicle are plotted in Figure 8.3 and Figure 8.4. The lateral states θ and $\dot{\psi}$ are tracked against the desired values. The other lateral state $\dot{\theta}$ is expected to approach to zero at steady state and V_y is expected to be stable and within limits. Similarly, all the states of the longitudinal dynamics are tracked against their respective desired signals. It can be observed that the desired states are being tracked successfully.

Figure 8.5 shows the plot of the lateral tyre forces of the NTTW vehicle while cornering. The front left and right tyre forces are almost equal and the rear tyre force is higher. The front left and right longitudinal tyre forces are shown in Figure 8.6, the rear tyre force is a purely rolling tyre and hence the force is zero.

The plot, Figure 8.7 showcases one of the important responses of the NTTW vehicle dynamics while cornering. The vertical forces represent road-holding capability of the vehicle. The tyre slip angles of the front left, right and rear wheels are shown in Figure 8.8. The ideal values should be as low as possible than the actual steering angle. Similarly, Figure 8.9 shows the plots of the slip ratios of the front left and right wheels of the NTTW vehicle during the test maneuver.

The plots of longitudinal, lateral and perceived accelerations are shown in the Figure 8.10. The integrated (lateral-tilt and longitudinal) desired path is compared with the actual path in Figure 8.11. It can be observed that even though the desired trajectory is quite steep, the performance of the IM-SM controller is satisfactory. This trajectory tracking is a lot better than the one reported for MVTD controller in Figure 6.19.

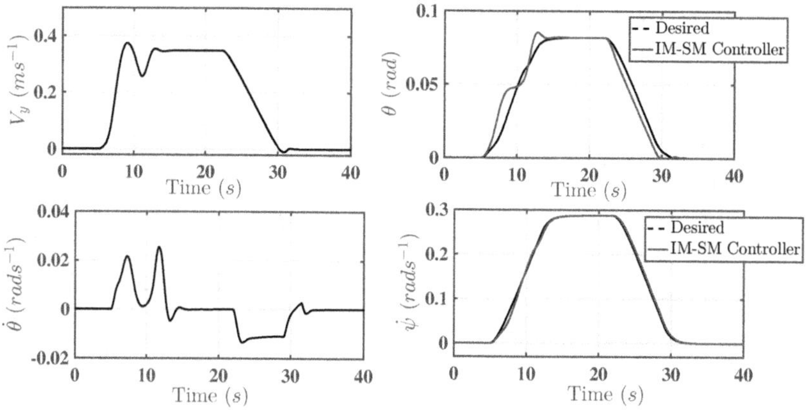

Figure 8.3: Lateral states for double lane change maneuver

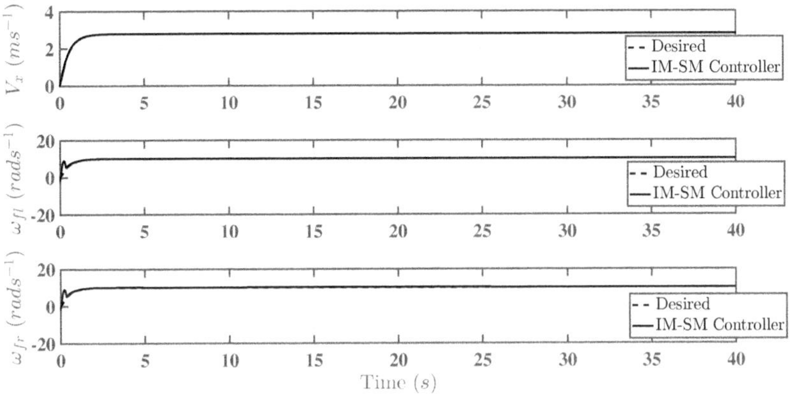

Figure 8.4: Longitudinal states for double lane change maneuver

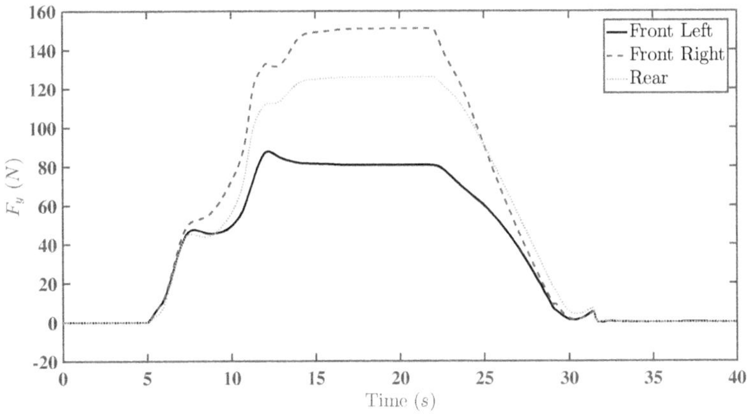

Figure 8.5: Lateral tyre forces for double lane change maneuver

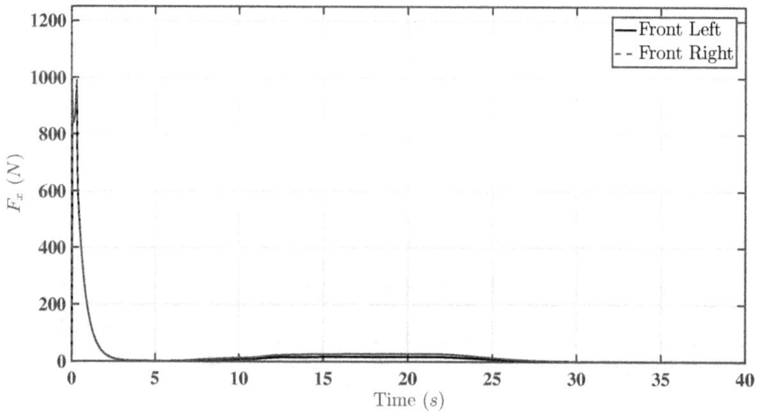

Figure 8.6: Longitudinal tyre forces for double lane change maneuver

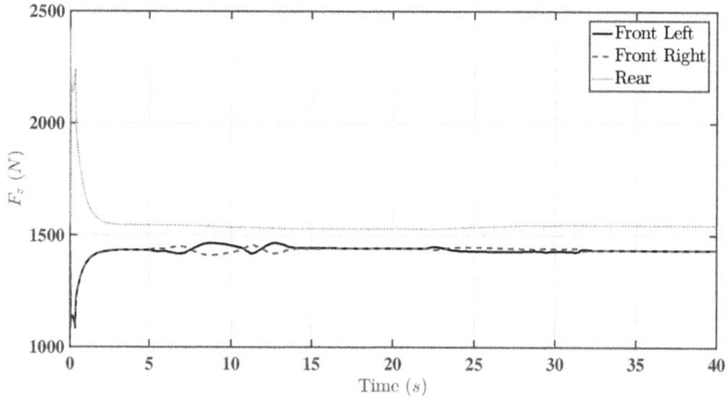

Figure 8.7: Vertical tyre forces for double lane change maneuver

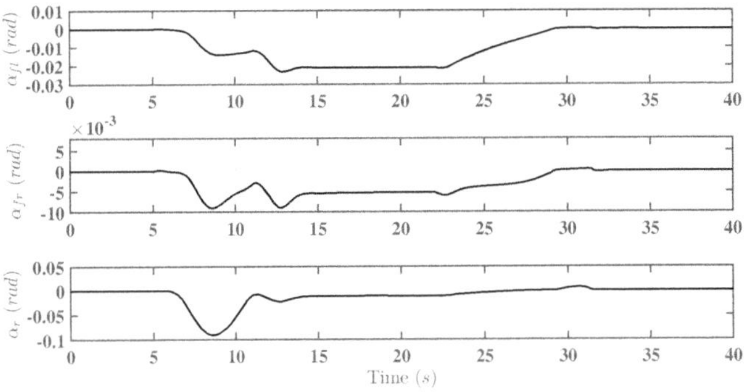

Figure 8.8: Tyre slip angle for double lane change maneuver

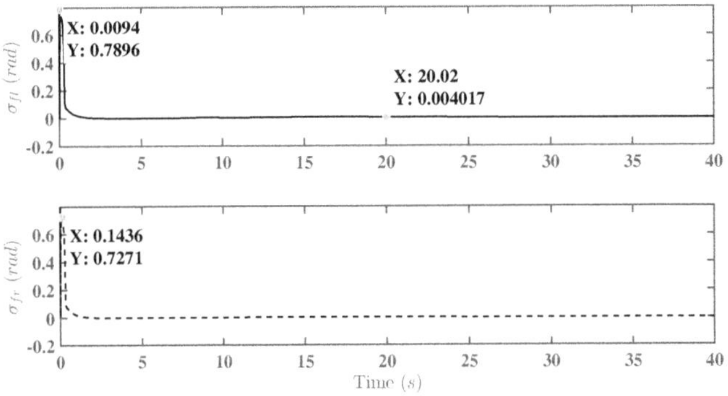

Figure 8.9: Tyre slip ratio for double lane change maneuver

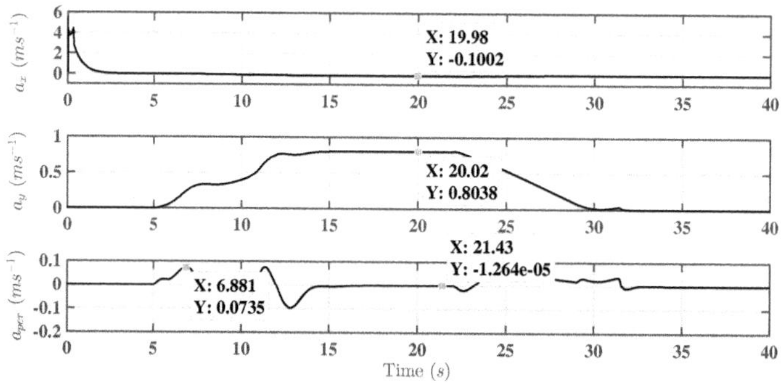

Figure 8.10: Different accelerations for double lane change maneuver

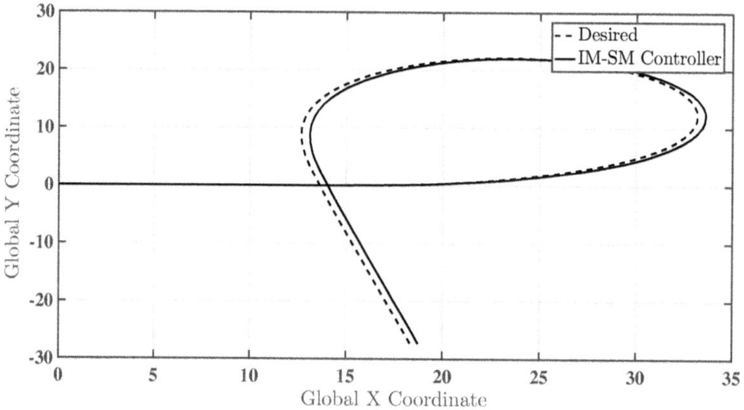

Figure 8.11: Vehicle trajectory for double lane change maneuver

8.4.2 Accelerating while cornering

A three wheeled vehicle is less stable as compared to a four wheeled vehicle due to lower trackwidth to CG ratio. Also, three wheeled vehicle has lower rollover speed limits while accelerating during cornering [5]. Hence, it becomes essential for evaluating the stability performance of NTTW vehicle while accelerating during cornering. The driver (speed command) accelerates from 0 to 20 kmph ($\approx 5.5ms^{-1}$) to 75 kmph ($\approx 21ms^{-1}$) while taking a constant radius of turn of fixed turning radius of curvature of 170 m. Figure 8.12 through Figure 8.16 represent the results obtained for this test case scenario. It can be observed from Figure 8.12 that during acceleration (20 to 35 s), the required control inputs increase and then settle when the constant steady state is achieved. Figure 8.13 represents the plot for all the lateral states. As expected, the vehicle tilts more as the speed increases from 0.01 rad (1 deg) at 20 kmph to 0.32 rad (18 deg) at 75 kmph. Figure 8.14 and 8.15 represent the perceived lateral acceleration and the trajectory of the NTTW vehicle during the maneuver. The Figure 8.16 shows the plot of the vertical tyre loading of the NTTW vehicle during cornering while acceleration. It can be observed that while acceleration (20 to 35 s), the vertical loading of rear wheel increases and simultaneously the loading at the front wheels decreases. This behavior is expected and it can be observed that the control inputs are modified to cope with this change.

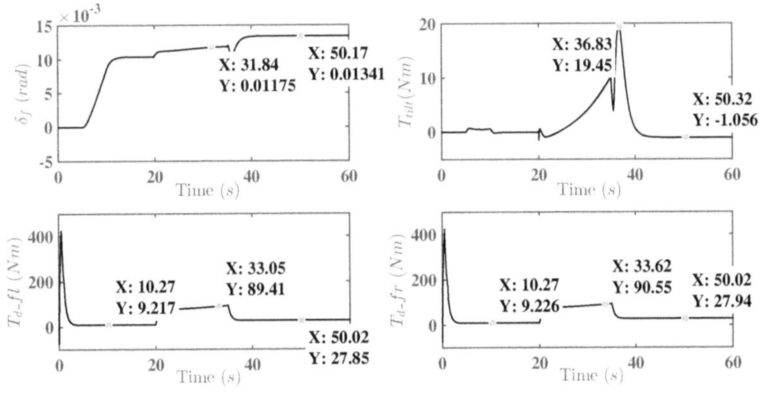

Figure 8.12: Control inputs during accelerating while cornering

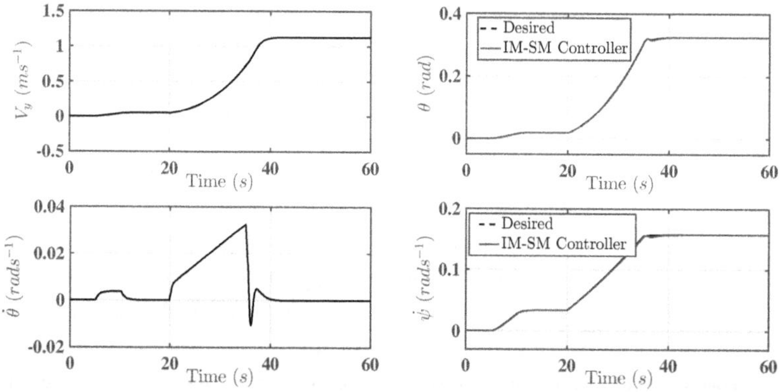

Figure 8.13: The dynamic states during accelerating while cornering

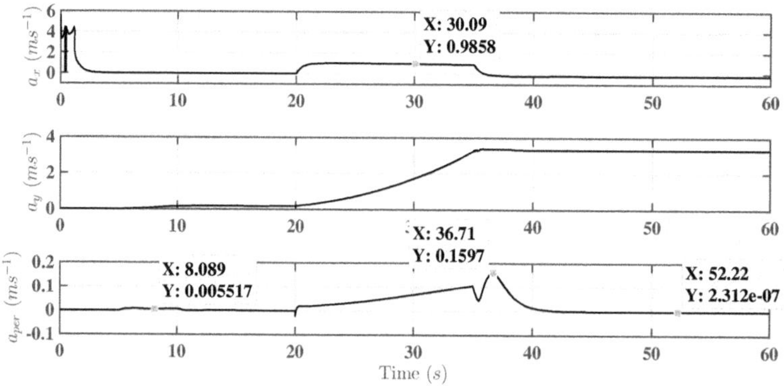

Figure 8.14: Perceived lateral acceleration during accelerating while cornering

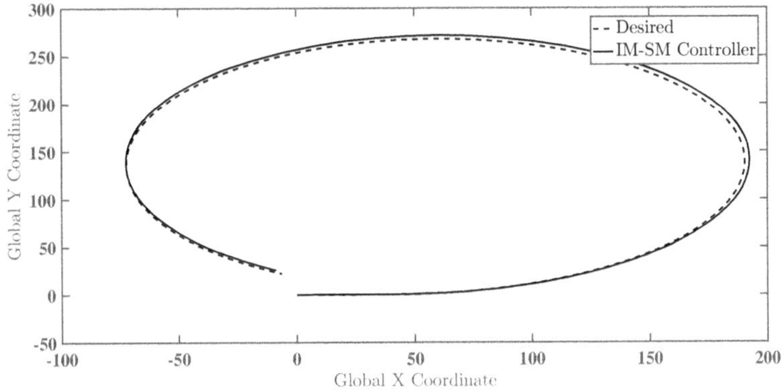

Figure 8.15: Trajectory during accelerating while cornering

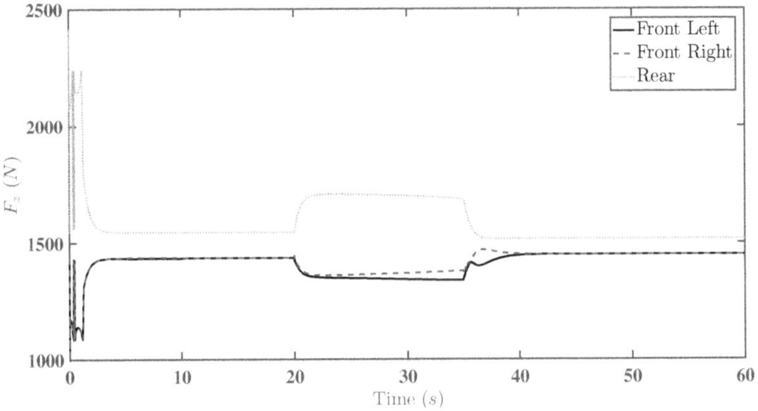

Figure 8.16: Vertical tyre loadings during accelerating while cornering

8.4.3 Cornering while deceleration

The braking system is not included in the scope of the thesis, hence a case of minimal deceleration is considered for analysis. The driver (speed command) accelerates from 0 to 30 kmph ($\approx 8ms^{-1}$)

and decelerates gradually to 25 kmph ($\approx 7ms^{-1}$) while taking a constant radius of turn of 135 m. Figure 8.17 through Figure 8.21 show plots for the control inputs, the vehicle states, body accelerations and vehicle trajectory. It can be observed from Figure 8.17 that during deceleration (10 to 50 s), the required control inputs increase and then settle when the constant steady state is achieved. Figure 8.18 represents the plot for all the lateral states. As expected, the vehicle tilting angle decreases as the speed of the vehicle decrease from 0.04 rad (2.5 deg) at 30 kmph to 0.037 rad (2.1 deg) at 25 kmph. Further, some peaks are observed during start of the deceleration period in the states. It is interesting to note that the drive torques also vary during deceleration. The trajectory plot shown in Figure 8.21 depicts the desired versus actual trajectory of the NTTW vehicle during cornering while deceleration.

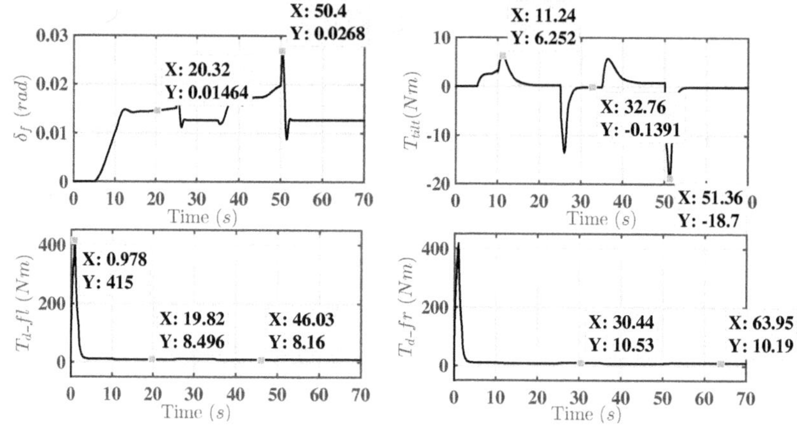

Figure 8.17: Control inputs for cornering while deceleration

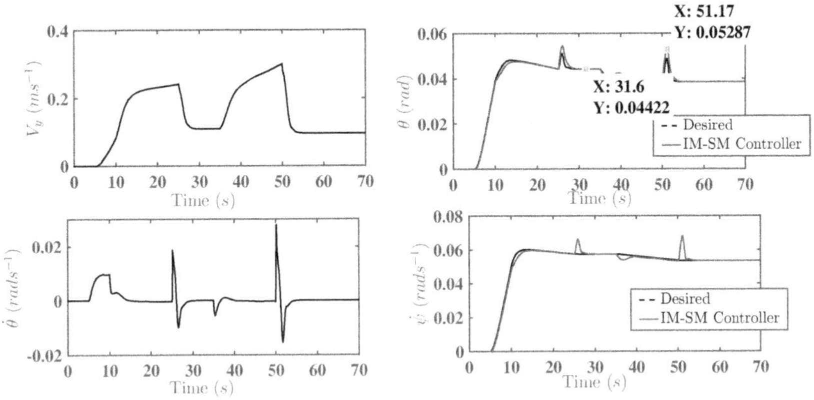

Figure 8.18: Lateral states for cornering while deceleration

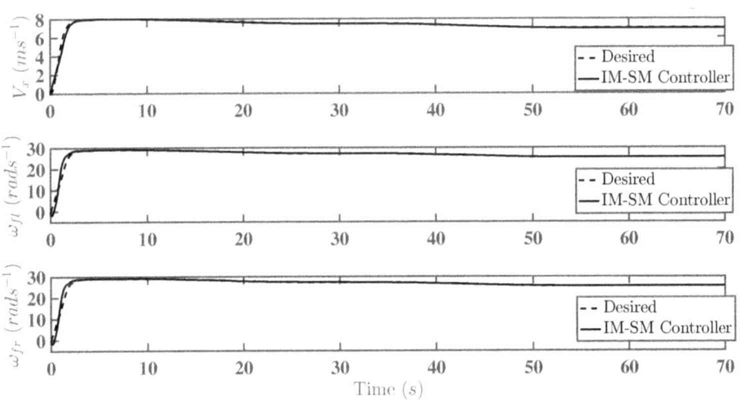

Figure 8.19: Longitudinal states for cornering while deceleration

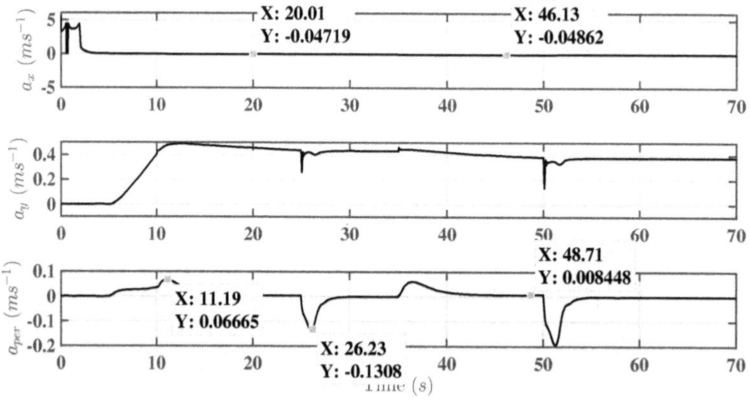

Figure 8.20: Body accelerations for cornering while deceleration

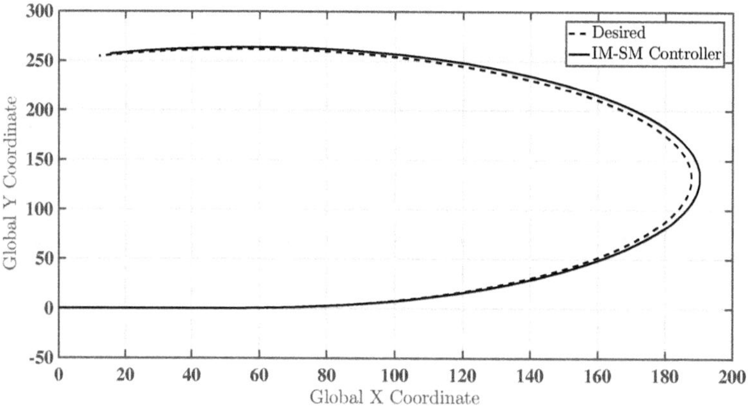

Figure 8.21: Vehicle trajectory for cornering while deceleration

8.5 Summary

i. A multivariable control design based on sliding mode control strategy (SMC) is presented for simultaneous control of the lateral-tilt states ($V_y, \dot{\theta}, \theta, \dot{\psi}$) with two inputs ($\delta_f$ and T_{tilt}). Similarly, the two drive torques of T_{d_fl} and T_{d_fr} are used to control all the longitudinal states (V_x, ω_{fl}, ω_{fr}) simultaneously. It is observed that the proposed controller (IM-SMC) and estimator (EKF) combination, is able to achieve the desired control objectives satisfactorily.

ii. Two separate controllers are designed for lateral-tilt and longitudinal dynamics as done in MVTD controller. Hence, the design of switching surface and the formulation of control law for SMC are also performed individually for each system. In addition, the system matrix of the lateral-tilt dynamics is modified to ascertain zero steady-state error tracking and thereby a modified SMC controller design is proposed. This formulation consisting of augmented integral states within standard SMC has been successfully implemented and shown to be meeting the stipulated control objectives. Based on the literature survey of various SMC applications to automotive chassis control such a technique seems to be unreported.

iii. The stability analysis of the proposed controller is performed to ensure that it facilitates finite time convergence of surface to zero. The stability is evaluated by employing Lyapunov analysis and the limiting conditions for gain parameters for the controllers are derived suitably based on Lyapunov criteria.

iv. Furthermore, the multivariable control solution proposed for lateral-tilt dynamics allows the usage of both the STC and DTC systems (SDTC), which basically means that the proposed strategies (MVTD and IM-SMC) can be seamlessly incorporated with the current technologies available in NTTW vehicles.

v. Also, it is observed that the drive torques obtained during simulations match well with analytical calculations given in Appendix A.

Chapter 9

Comparison of MVTD and IM-SM Controllers

9.1 Introduction

This thesis proposes two multivariable controllers for integrated dynamics of NTTW vehicle. It is necessary to discuss and bring out pros and cons of both the controllers to reach a final optimal controller design suitable for stability controller for NTTW vehicle. This chapter presents the detailed comparison of the two proposed multivariable controllers, i.e. MVTD and IM-SM controllers. The comparison of the baseline controller (Chapter 5) and the MVTD controller (Chapter 6) is presented in Section 6.4.1. It is concluded that the proposed MVTD controller is better than the baseline controller in terms of vehicle handling and stability performance.

In this chapter, a detailed comparison between the two proposed multivariable controllers, i.e. the MVTD controller and the IM-SM controller is presented. Different test scenarios have been simulated and the corresponding responses are monitored. A detailed investigation is showcased by means of quantitative, such as input power requirements, as well as qualitative analysis, such as tracking response. The responses of each controller are compared in presence of external disturbances such as variation in vehicle resistance forces, additive disturbance in input channel and variations in parametric values.

9.2 Step maneuver

A step maneuver with a constant speed of 40 kmph ($\approx 11.11ms^{-1}$) is simulated to test the controller. Initially, the vehicle moves in a straight line for 10 sec, then the vehicle takes a turn of 75 m radius of curvature. Figure 9.1 through Figure 9.4 show the plots that compare the control inputs, controlled states, perceived lateral acceleration and the trajectory of the NTTW vehicle with MVTD and the IM-SM controllers. The 'Desired' signals depicted in the plots are generated by the reference model as presented in Section 4.3. It can be observed from Figure 9.1 and Figure 9.3 that apart from some variation in transient response, the steady-state values are similar for both the strategies. Moreover, when the control efforts of both the strategies are compared (in 9.1), the δ_f is higher in IM-SM controller (0.03 rad) than MVTD controller (0.025 rad). This is reflected in the state dynamics plot in Figure 9.2 where it can be observed that the IM-SM controller follows the desired trajectory better. However, more clarity is obtained by observing the trajectory of the NTTW vehicle, viz. in Figure 9.4. It can be inferred that the desired trajectory is being followed more closely by the IM-SM controller than the MVTD controller.

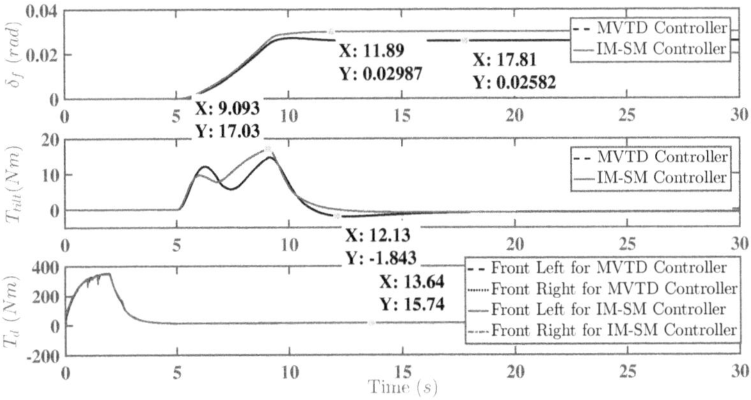

Figure 9.1: Control inputs of the two strategies

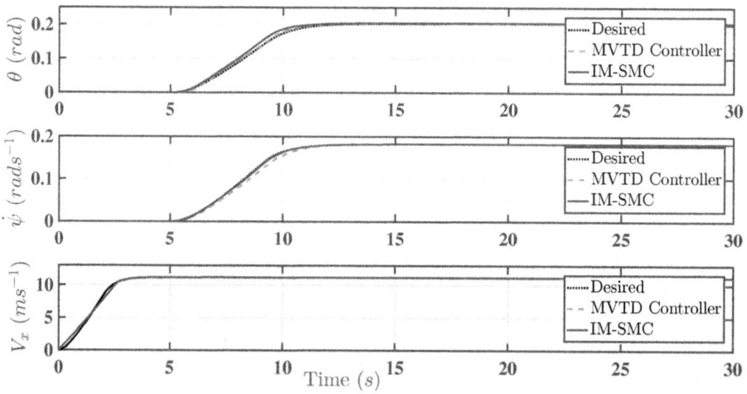

Figure 9.2: Dynamic states of the two strategies

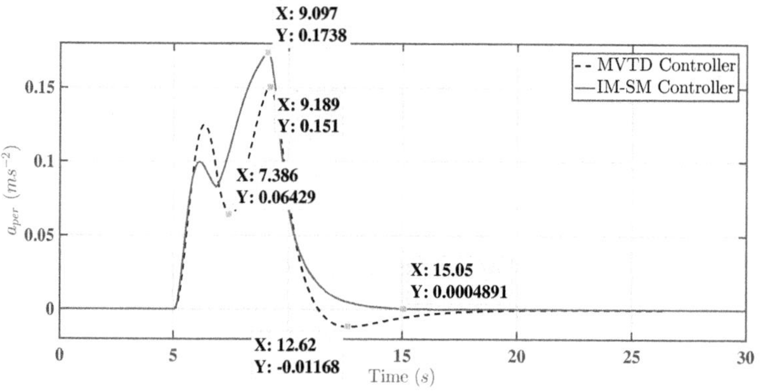

Figure 9.3: Perceived lateral accelerations of the two strategies

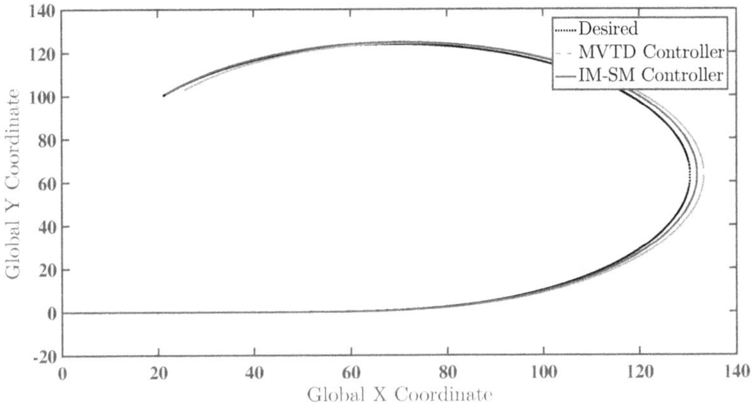

Figure 9.4: XY trajectories of the two strategies

9.3 Double lane change maneuver

A double lane change maneuver with around 23 m of turning radius and constant speed of 20 kmph ($\approx 5.5ms^{-1}$) is simulated to compare the two multivariable control strategies. Initially, the vehicle moves in a straight line for 5 sec. Then, the driver takes a turn of 23 m radius of curvature. At 20 seconds, the driver changes his direction of travel to a straight line till the end of the simulation. Figure 9.5 through Figure 9.8 show the plots that compare the control inputs, controlled states, perceived lateral acceleration and the trajectory of the NTTW vehicle with MVTD and the IM-SM controllers. This test case provides a more clear picture with respect to the transient response of each of the controller. It can be observed in Figure 9.5 and Figure 9.7 that there is more variation in the T_{tilt} (varies from 9.8 Nm to -12.22 Nm) and the a_{per} (varies from 0.1 ms^{-2} to -0.1 ms^{-2}) during the transient period with the MVTD controller. Whereas the performance of the IM-SM controller during transients is smoother (T_{tilt} varies from 9.8 Nm to -6.3 Nm and a_{per} varies from 0.1 ms^{-2} to -0.05 ms^{-2}). This is one of the advantages of the IM-SM controller as smoother transient behavior would lead to better handling and comfort for the driver. Further, the desired trajectory is also followed more closely as can be observed from Figure 9.8.

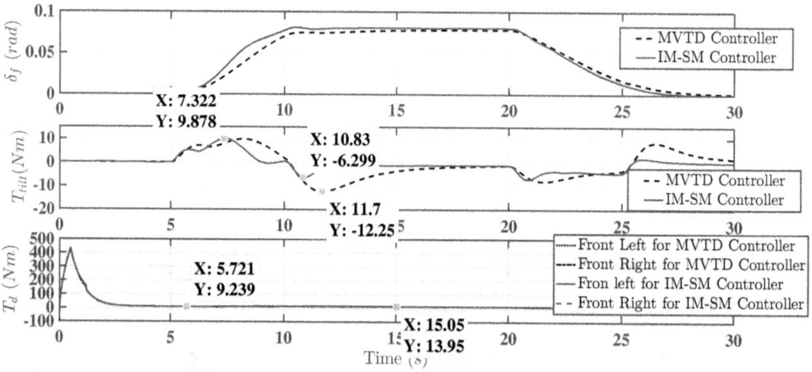

Figure 9.5: Control inputs of the two strategies

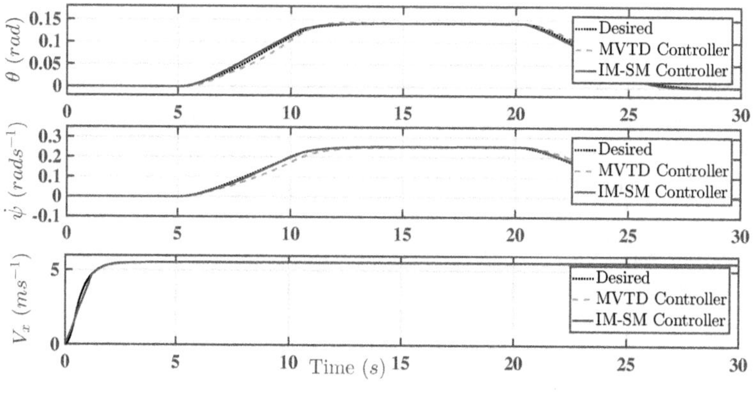

Figure 9.6: Dynamic states of the two strategies

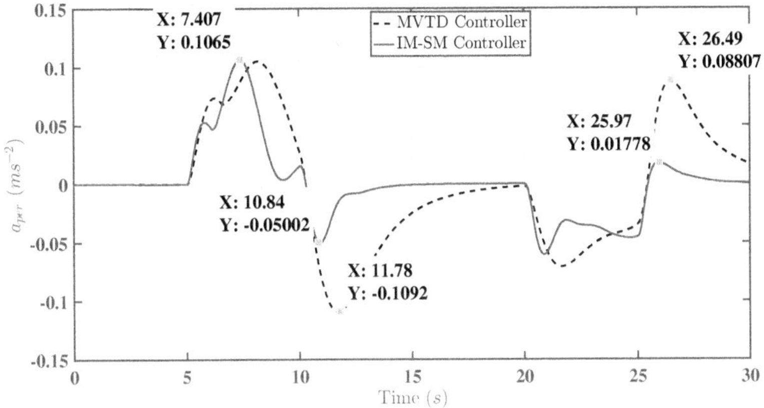

Figure 9.7: Perceived lateral accelerations of the two strategies

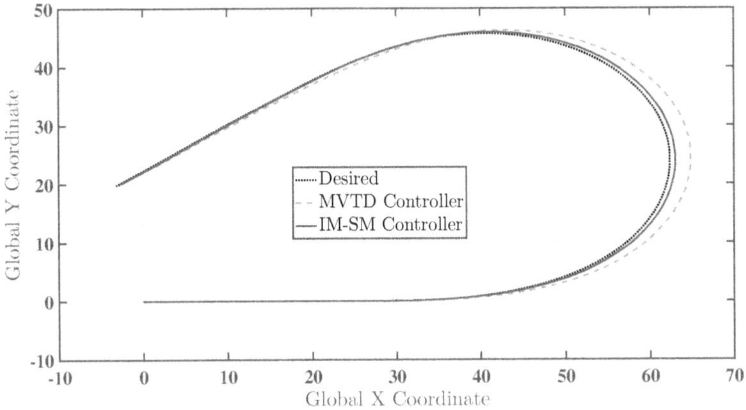

Figure 9.8: XY trajectories of the two strategies

9.4 Double lane change maneuver with Disturbances

The same maneuver (double lane change maneuver with around 23 m of turning radius and constant speed of 20 kmph) as used in Section 9.3 is further used to analyze the performances of the controllers in presence of different types of external disturbances. The MVTD and IM-SM controllers are compared in presence of the disturbances as listed in Table 4.5 individually. Figure 9.9 through Figure 9.20 in subsequent sub-sections show the plots which compare the control inputs, controlled states, perceived lateral acceleration, and trajectories.

9.4.1 Double lane change maneuver with input disturbance

Figure 9.9 through 9.12 show the comparison plots of control inputs, controlled states, perceived lateral acceleration and the trajectory of the NTTW vehicle in presence of constant input matched disturbance. 20% of maximum inputs during the maneuver is added at the respective inputs (δ_f and T_{tilt}) to emulate matched input disturbance. The disturbance is added at 14 s from the start of the simulation and hence, the changes can be observed after this point of time. It can be observed from Figure 9.9 that the controllers readjust the values of the control inputs to reject the effect of external disturbances on the vehicle dynamics and thus to provide the necessary performance. The maximum values of control inputs (Figure 9.9) are almost same for both the controllers, however, the variation in the transient responses are more for MVTD controller similar to the observation made in previous sections. Moreover, due to the addition of disturbance at 14 s, the demand of motor drive torque has increased (from 14 Nm in Figure 9.5 to 19 Nm in Figure 9.9). The desired states are followed closely by both the strategies with a slight peak at the point where disturbance is added as shown in Figure 9.10. It is interesting to notice that the value of θ decreases (0.14 rad to 0.13 rad) in IM-SM controller at 14 s, whereas the value is increased (0.14 rad to 0.15 rad) in case of MVTD controller. Another observation can be made from the plot of a_{per} in Figure 9.11 that the response of IM-SM reaches the peak faster(7.3 s) than the MVTD controller (8 s) and similar observation can be made from Figure 9.9. The trajectory response, as expected, indicates that the IM-SM controller is better at following the desired trajectory than the MVTD controller.

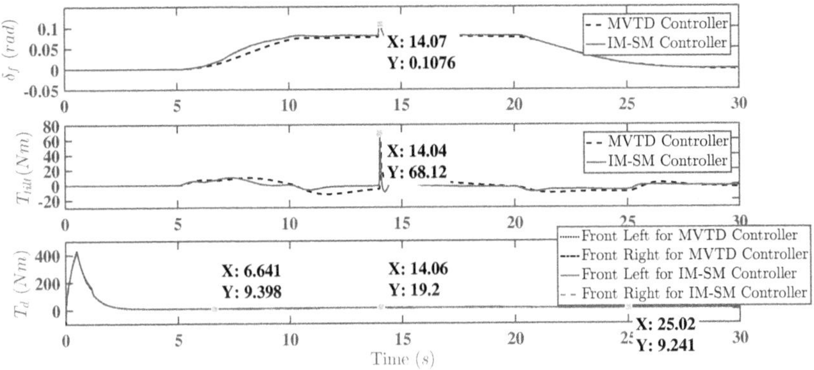

Figure 9.9: Control inputs of the two strategies

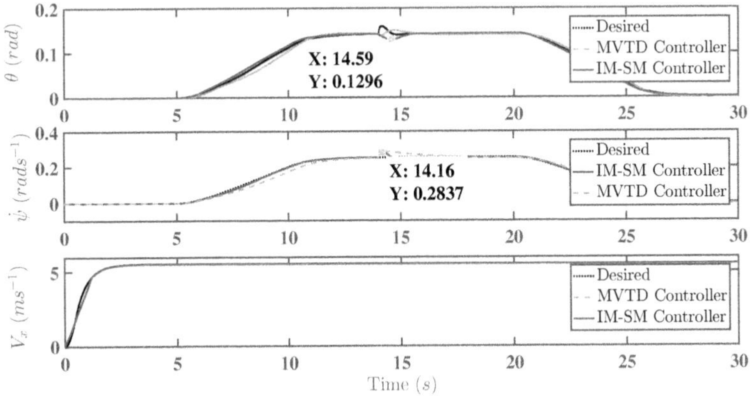

Figure 9.10: Dynamic states of the two strategies

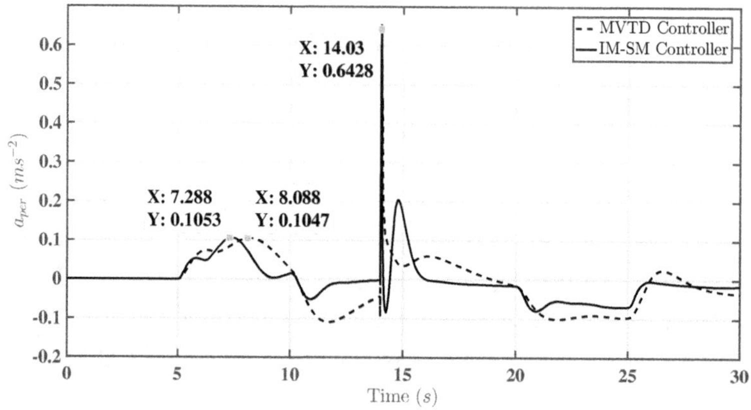

Figure 9.11: Perceived lateral accelerations of the two strategies

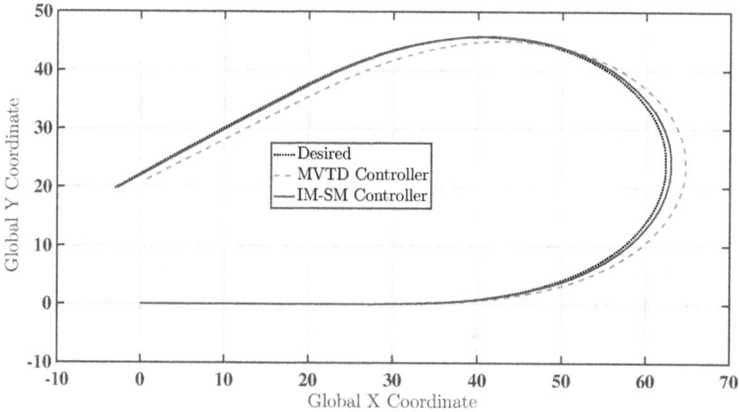

Figure 9.12: XY trajectories of the two strategies

9.4.2 Double lane change maneuver with Longitudinal drag

Figure 9.13 through Figure 9.16 show the plots comparing the performance of the proposed multivariable controllers when a sinusoidal disturbance is added in the longitudinal drag component of the longitudinal dynamics. 5% of maximum F_x is added to the F_{aero} component at 14 s of simulation. In Figure 9.13 and Figure 9.15 more peak is observed in T_{tilt} and a_{per} respectively in MVTD controller due to the addition of longitudinal drag disturbance. The peak due to disturbance addition in T_{tilt} is higher in MVTD (-38 Nm) than in IM-SM (-25 Nm) controller. Similarly, peak and variation in a_{per} are also higher for MVTD controller (0.08 ms^{-2} to -0.2 ms^{-2}) than in IM-SM controller (0.1 ms^{-2} to -0.05 ms^{-2}) as observed in Figure 9.15. Similar to the previous section, it is observed that the steady state values of the desired states are achieved well equally by both the strategies with a slight peak at the point of addition of disturbance (Figure 9.14. The desired trajectory is followed in a better way by the IM-SM controller.

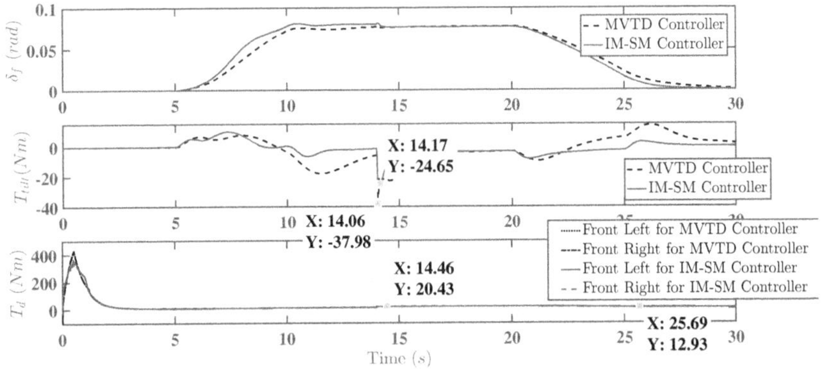

Figure 9.13: Control inputs of the two strategies

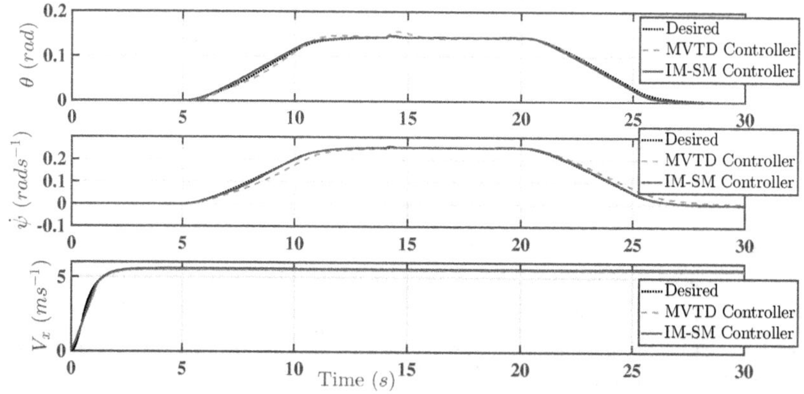

Figure 9.14: Dynamic states of the two strategies

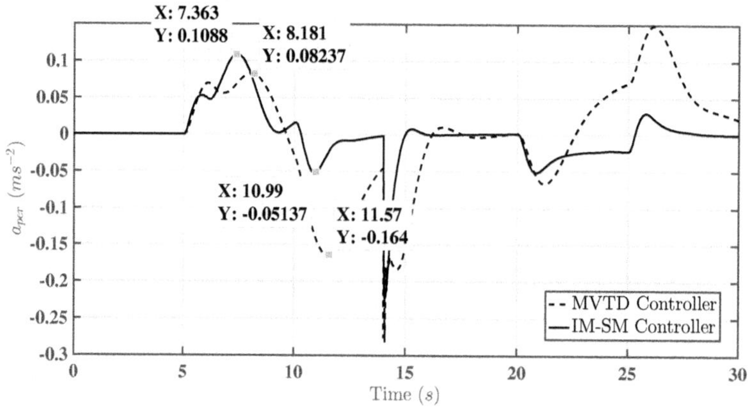

Figure 9.15: Perceived lateral accelerations of the two strategies

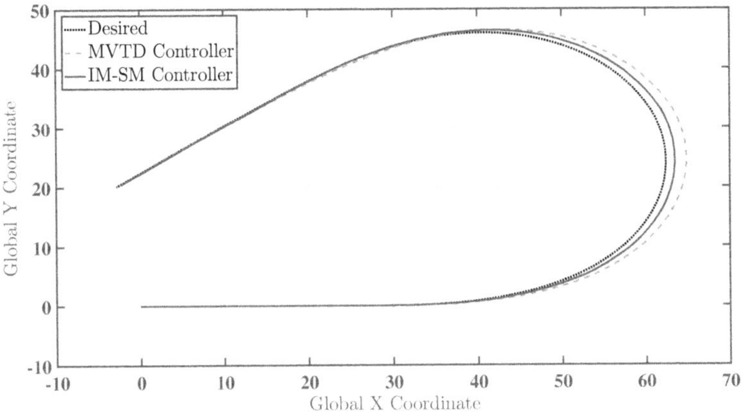

Figure 9.16: XY trajectories of the two strategies

9.4.3 Double lane change maneuver with gradient

Another interesting and practical disturbance is consideration of road gradient during a maneuver, which would essentially mean an increase in road resistance force. 5 deg of road gradient is considered to be present from 14 s to 18 s which consequently adds external force at F_{grad}. Figure 9.17 through Figure 9.20 show the plots comparing the performance of the proposed multivariable controllers in presence of the above mentioned disturbance. The plots show similar responses as observed in earlier sections. The transient response is smoother and the desired trajectory os followed more closely in case of IM-SM controller and thus, it can be inferred that the proposed multivariable IM-SM controller performs better than the MVTD controller. The results also indicate that due to the coupling between the lateral and longitudinal dynamics through tyres in the NTTW vehicle model, addition of external disturbances on one dynamics would influence the other as well.

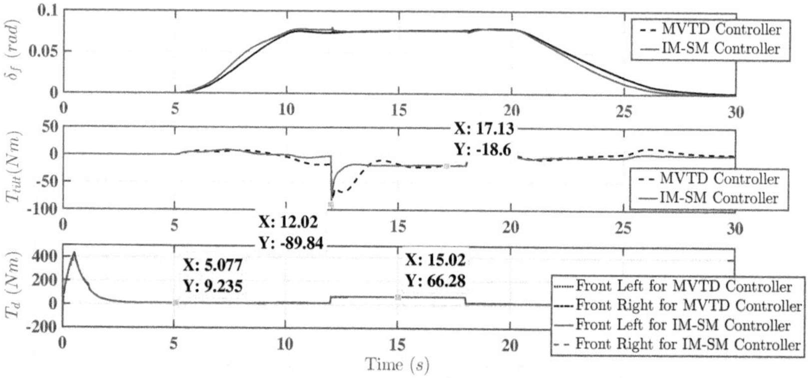

Figure 9.17: Control inputs of the two strategies

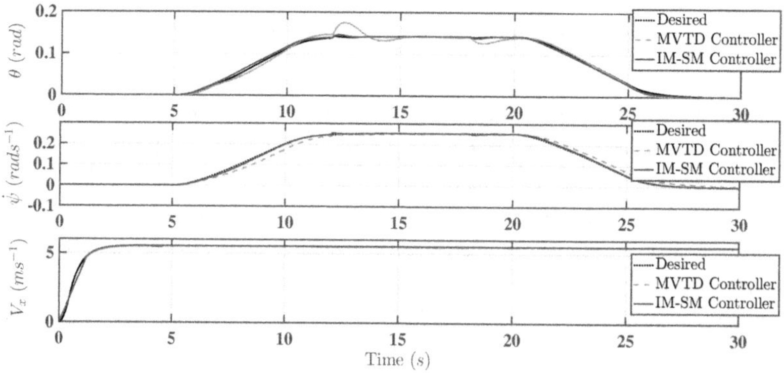

Figure 9.18: Dynamic states of the two strategies

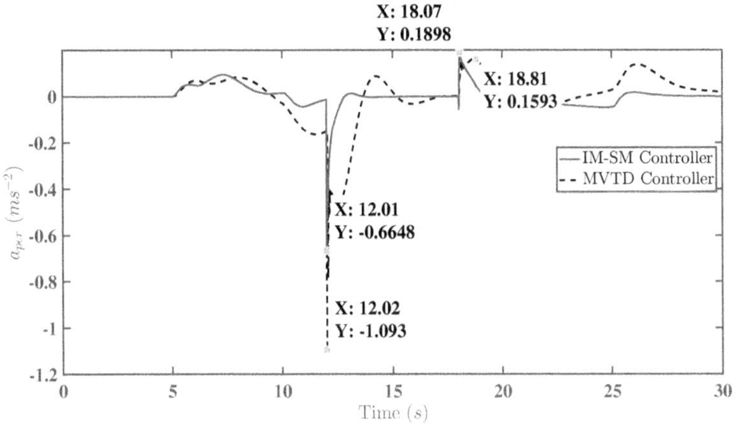

Figure 9.19: Perceived lateral accelerations of the two strategies

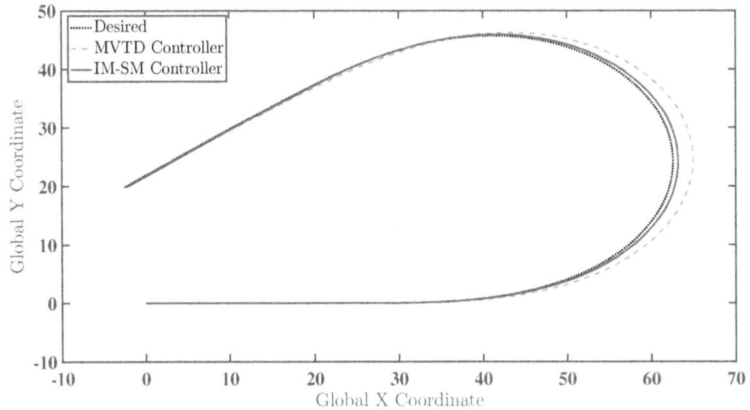

Figure 9.20: XY trajectories of the two strategies

9.5 Discussion

In this section, discussion on analysis has been performed in qualitative and quantitative terms to compare the performance of MVTD and IM-SM controllers.

9.5.1 Qualitative analysis

The qualitative analysis of the two controllers is performed by studying the transient phase of the control inputs and the perceived lateral acceleration. It can be observed from the plots of a_{per} shown in Figures 9.3, 9.7, 9.11, 9.15 and 9.19 that the variation in its response is larger in MVTD controller than in IM-SM controller. Similar behavior is noticed in the control inputs shown in Figures 9.1, 9.5, 9.9, 9.13 and 9.17, where the input peaks are lower and the transient phase is smoother in case of IM-SM controller as compared to MVTD controller. This is one of the advantages of the IM-SM controller as smoother transient behavior would lead to better handling and comfort for the driver. Moreover, it can be inferred from the test maneuvers reported in this chapter that the dynamic states of the NTTW vehicle follow the desired command of the driver more closely than the MVTD controller where the transient time of the states are slower (Figures 9.2, 9.6, 9.10, 9.14 and 9.18). This is even more evident in the plots representing the XY trajectories of the two strategies. The trajectory plots shown by 9.21 and Figure 9.22 give a more clear view of the effect of the disturbances on the evolution of the states of the NTTW vehicle. The desired values shown in the plots represent the original desired value when no disturbance is present. As a disturbance is added, some minor changes in the desired value are observed and the trajectories follow the respective desired values well. However, the trajectory plots shown by 9.21 and Figure 9.22 compare the respective (disturbance) trajectories with the original trajectory (without disturbance) for MVTD and IM-SM respectively and hence, slight deviation is observed in some cases. It is observed that the desired trajectory is followed quite well in case of matched input disturbances, however, a slight deviation is observed from the originally desired trajectory in case of the longitudinal disturbances. Nevertheless, the NTTW vehicle is stable throughout the maneuver which shows the IM-SMC performs satisfactorily in presence of disturbances. The conclusive points for qualitative analysis can be listed as:

1. IM-SM controller is able to achieve the desired trajectory faster
2. Due to less variation in a_{per}, the ride is more comfortable for the driver in case of IM-SM controller
3. The maximum trajectory error is lesser for IM-SM controller than that for MVTD controller

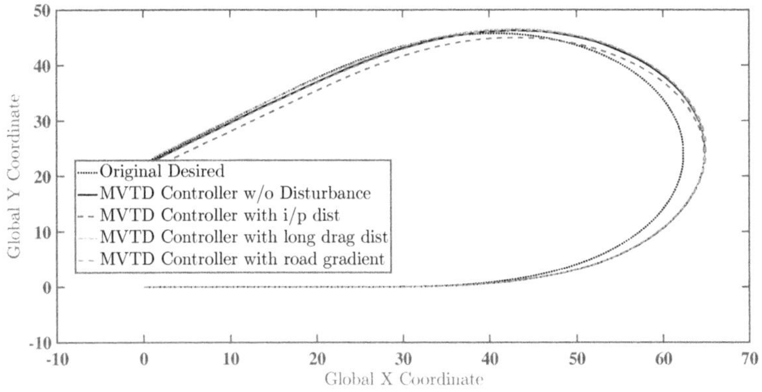

Figure 9.21: XY trajectories of MVTD controller

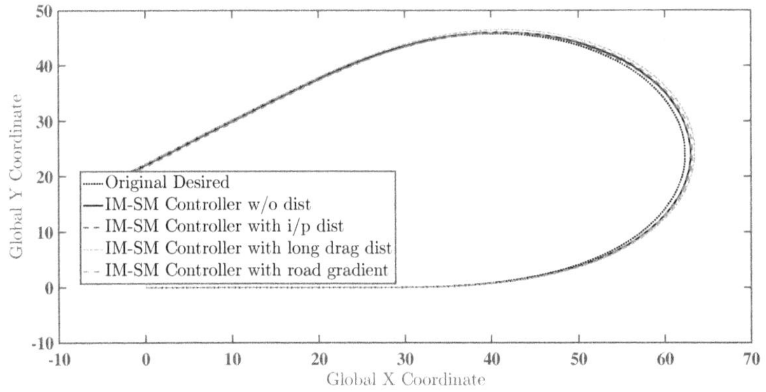

Figure 9.22: XY trajectories of IM-SM controller

9.5.2 Quantitative analysis

To evaluate the actual difference between the control efforts (U) of the MVTD controller and IM-SM controller in quantitative terms, power utilization of the control efforts is calculated during different maneuvers. The mathematical expression for the equivalent power is given by (6.6).

Table 9.1 provides a comparison between the control input powers of IM-SM and MVTD controllers during different test scenarios reported in this chapter. The total input power captures the amount of energy utilization by all the four input actuators, i.e. steer-by-wire, tilt and two in-wheel drive actuators during the simulation period. It is observed from Table 9.1 that for step maneuver the MVTD controller utilizes 0.71% less power than that of IM-SM controller. However, in all other test scenarios, the IM-SM controller draws less power than MVTD controller. Moreover, as external disturbances are added, the power utilization increases for each controller. This is expected as the controller needs to modify the input efforts in order to reject the effect of disturbances. There is an appreciable difference of 5.6% in the power requirement between MVTD and IM-SM controller for double lane change maneuver when external disturbance of longitudinal drag is added. Furthermore, if the original power of double lane change maneuver is compared with the one with added disturbance for IM-SM controller, a power increase of 3% is noticed. Hence, it can be inferred that the input power requirements of the controllers are comparable in most scenarios. However, the MVTD requires more input power in the presence of external disturbances such as variation in vehicle resistance forces.

Table 9.1: Control input power

Control Inputs ⇒		Total Power	% Difference
Step maneuver	$MVTD$	14137	-0.71
	$IM-SM$	14237	
Double lane change maneuver	$MVTD$	6682	0.015
	$IM-SM$	6681	
Double lane change maneuver with input dist	$MVTD$	6700	0.05
	$IM-SM$	6697	
Double lane change maneuver with long drag	$MVTD$	6876	5.6
	$IM-SM$	6493	
Double lane change maneuver with gradient	$MVTD$	8578	1.5
	$IM-SM$	8451	

Table 9.2 tabulates the comparison between the state errors of MVTD and IM-SM controller when parameter uncertainty is introduced in the system. The data is collected for double lane change maneuver with 20 kmph constant velocity and 23 m radius of turn. Maximum tracking error of respective vehicle states (θ, $\dot{\psi}$ and ψ) are compared when parameter variations of 20% are added in m_{cg}, I_{z_veh} and I_{x_veh}. It can be observed that the state errors are comparatively higher

in case of MVTD controller than the IM-SM controller.

Table 9.2: State error with parameter uncertainty for double lane change maneuver

State error ⇒		$\lvert\theta_{e_{max}}\rvert$ (rad)	$\lvert\dot{\psi}_{e_{max}}\rvert$ $(rads^-1)$	$\lvert\psi_{e_{max}}\rvert$ (rad)
Original	$MVTD$	6×10^{-3}	0.0273	0.134
	$IM-SM$	5×10^{-3}	7.5×10^{-3}	0.0236
20% uncertainty in m_{cg}	$MVTD$	7.1×10^{-3}	0.0236	0.132
	$IM-SM$	4×10^{-3}	4.9×10^{-3}	0.022
20% uncertainty in I_{z_veh}	$MVTD$	6×10^{-3}	0.0288	0.1314
	$IM-SM$	5.65×10^{-3}	8.6×10^{-3}	0.0236
20% uncertainty in I_{x_veh}	$MVTD$	6×10^{-3}	0.0273	0.134
	$IM-SM$	4.9×10^{-3}	7.5×10^{-3}	0.0236

Similarly, Table 9.3 tabulates the comparison between the time required to reject the external disturbances by MVTD and IM-SM controller. As mentioned above, the data is collected for double lane change maneuver with 20 kmph constant velocity and 23 m radius of turn. The time required to reject disturbance is mapped from the plots of θ (T_θ) and $\dot{\psi}$ ($T_{\dot{\psi}}$). It can be observed that the MVTD controller takes more time to reject the external disturbances than the IM-SM controller.

Table 9.3: Time response for disturbance rejection

Time for disturbance rejection ⇒		T_θ (s)	$T_{\dot{\psi}}$ (s)
Double lane change maneuver with input dist	$MVTD$	1.57	2.31
	$IM-SM$	1.57	0.66
Double lane change maneuver with long drag	$MVTD$	2.18	0.53
	$IM-SM$	0.62	0.53
Double lane change maneuver with gradient	$MVTD$	3.21	2.8
	$IM-SM$	1.06	0.51

9.6 Summary

i. A detailed analysis is performed to compare the proposed multivariable integrated controllers in qualitative and quantitative terms. The comparison is made between proposed integrated multivariable stability controllers for NTTW vehicle using different test cases. It is observed

Chapter 10

Hardware-In-the-Loop (HIL) Testing

10.1 Introduction

The simulation results illustrated in the previous chapters are satisfactory in the sense that control objectives for vehicle stability are accomplished. However, further verification of the control is required in the real time. Hardware-in-the-loop (HIL) simulation has increasingly become popular for testing and validating a control system design. This kind of set up allows evaluating the components of the system which are hard to model or which are of high cost to build. Real-time interaction between simulation and hardware component enables the study of the control system in a more practical way. Many functions of HIL simulation platform include testing of the hardware performance of vehicle control system, the stability of communication and the logic errors of vehicle control strategy.

In the current scenario, desktop computers are powerful enough to run a basic dynamic HIL simulation with a tolerable error. In this work, an HIL simulation platform is built for NTTW vehicle control system validation. This chapter first presents the various HIL platforms built in literature, followed by a detailed explanation of the set up that is designed and developed in this work. The HIL simulation results for each of the controllers are presented and discussed.

10.2 HIL set up for this work

In this work, hardware-in-the-loop set up is designed and built to test and validate the stability control system of an electric/ hybrid-electric NTTW vehicle. The configuration of the set-up developed for this work is shown in Figure 10.1. The HIL comprises of a host machine (simulating the NTTW vehicle model) and a target processor (housing the controller algorithm), which communicate with each other using suitable data acquisition box. This interaction between the host and the target enables the study of the control system in a more practical way. The host is a Microsoft Windows based personal computer that contains the NI PCIe-6353 board and supporting software, such as MatLab/SimuLink of Mathworks, STM32 ST-link Utility driver. The NI breakout board serves as an interface between the host computer and the target board by assigning channel numbers to the respective boards and converting the electric signals into physical numbers. The input and output signals from the host computer and target board are connected through the NI data acquisition system (DAQ). The target board is a low power ARM based 32-Bit ARM Cortex-M4 CPU termed as STM32F. The STM32F board is used as target board on which the control algorithm resides. The controllers are programmed on the target board via STM link. This HIL set up is used to validate that the proposed algorithms can be deployed on the embedded systems successfully.

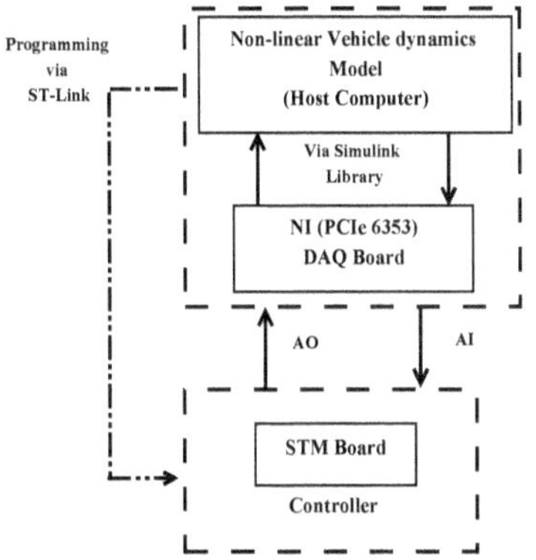

Figure 10.1: Configuration of the hardware-in-the-loop set up used in this work

The complete HIL set-up consists of hardware components and software to execute the required simulation. The subsequent sections document the details of the complete HIL simulation set-up. The hardware-in-the-loop system consists of the host computer, the target board and the interface board. The lab set up used in this work is shown in Figure 10.2.

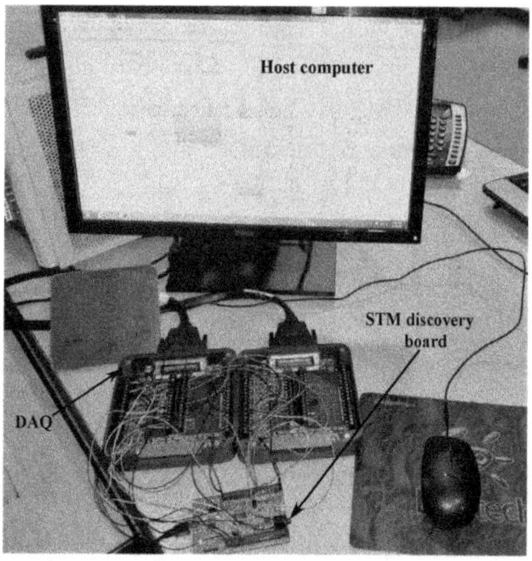

Figure 10.2: Lab set up

10.2.1 Control Algorithm

The control block contains the chassis dynamics stability control algorithms designed in this work for the NTTW vehicle. This block is open to the users so that they can modify and test the various control algorithms under different test cases. The control algorithm consists of multiple Simulink blocks which are compiled into code using Waijung library blocks and programmed on the target board via STM link. All three controllers reported in this thesis work described in Chapter 5 (control described in (5.1) and (5.2)), Chapter 6 (control described in (6.4)) and Chapter 8 (control described in (8.15)) have been programmed on the target board to test the efficacy of each controller design when implemented within an embedded platform.

10.2.2 Vehicle model

The Simulink Desktop Real Time in the host machine executes the non-linear mathematical model of the NTTW vehicle. Thus, the Simulink model represents the characteristics of the nonlinear dynamics of the NTTW vehicle. This provides flexibility for modifications in the mathematical model. The non-linear mathematical model of NTTW vehicle described in Chapter 3 in (3.17) to (3.23) is programmed in the host computer.

10.3 Results and Discussion

In this section, results of the HIL simulation are presented to validate all the proposed three controllers, viz. baseline, MVTD and IM-SM Controllers. The HIL results and offline simulation results of each controller are compared in each of the plots. Figure 10.3 through Figure 10.8 represent the plots of baseline, MVTD and IM-SM controllers respectively using step change and double lane change maneuver. For the step change maneuver, initially, the vehicle moves in a straight line for 5 sec. Then, the driver takes a 28 m of turn radius of curvature from 5 to 10 seconds and continues until the end of the simulation. Similarly, a double lane change maneuver of 50 m turn radius and speed of 35 kmph ($\approx 10ms^{-1}$) is simulated. The complete baseline controller (described in Chapter 5), i.e. lateral-tilt as well as longitudinal dynamics controller is programmed on the target board. It can be observed from the Figure 10.3 and Figure 10.4 that the HIL and offline simulation results match well with each other. However, small jitters can be observed in the response of HIL simulation. This may be due to the resolution mismatch between NI board (16-bit) and target board (12-bit). Quantization error of states and control inputs due to ADC and DAC are given in Table 10.1 and Table 10.2. It can be observed that the values of the quantization errors are not perceivable with respective to the maximum values of the respective states and control inputs and hence do not affect the performance substantially.

Table 10.1: Quantization error of states

States	Quantization error
V_y	$5 \times 10^{-4}\ ms^{-1}$
$\dot{\theta}$	$2 \times 10^{-4}\ rads^{-1}$
θ	$2 \times 10^{-4}\ rad$
$\dot{\psi}$	$4 \times 10^{-4}\ rads^{-1}$
V_x	$9.8 \times 10^{-3}\ rads^{-1}$

Table 10.2: Quantization error of states

Control inputs	Quantization error
δ_f	2.5×10^{-4} rad
T_{tilt}	0.015 Nm
T_{d_max}	0.15 Nm
T_{d_av}	0.03 Nm

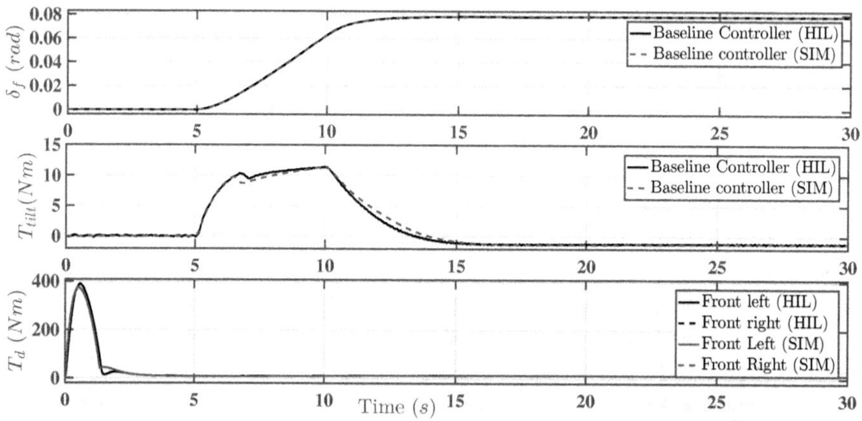

(a) Control inputs

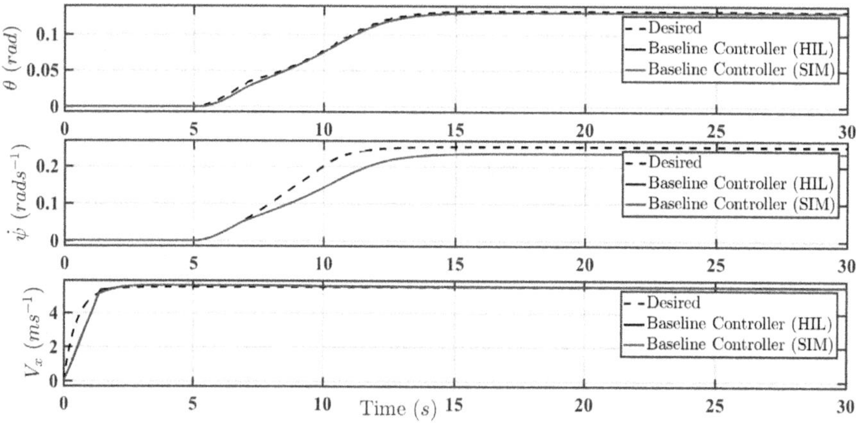

(b) States

Figure 10.3: HIL simulation results for baseline controller during a step change

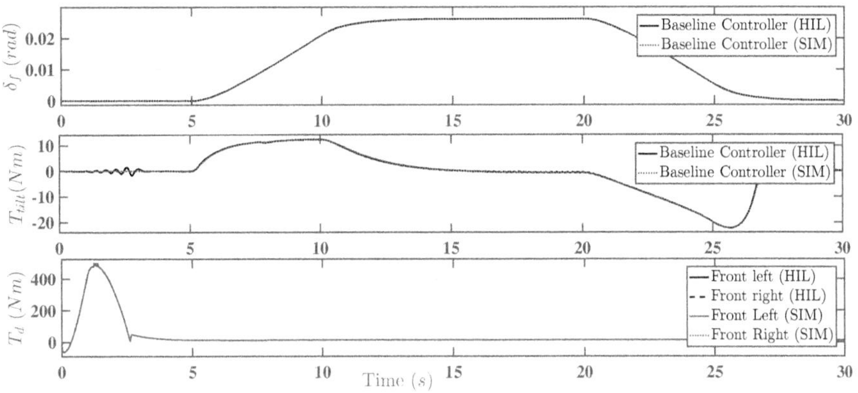

(a) Control inputs

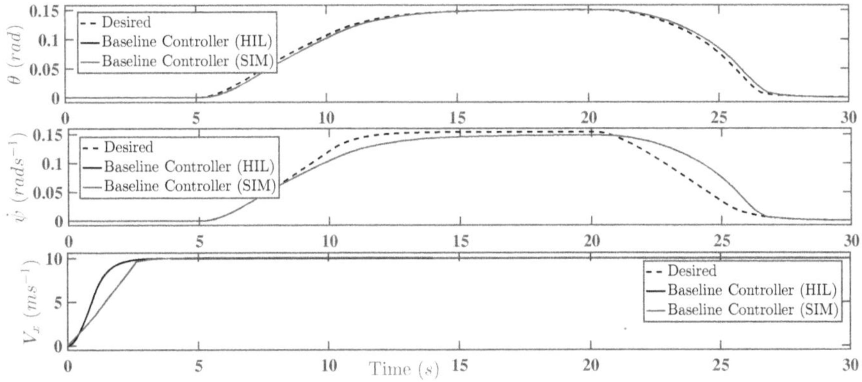

(b) States

Figure 10.4: HIL simulation results for baseline controller during double lane change

Similar maneuvers are simulated for MVTD (lateral-tilt) controller as described in (6.4) and the HIL and offline simulations are compared. It can be observed from the Figure 10.5 that the HIL and offline simulation results are matching. Similar to the previous plot of baseline controller, some oscillations are observed in the control inputs of HIL simulation. The jitters have increased to some extent which may be due to the fact that only the lateral-tilt controller is programmed in the target block. Now as there is coupling between the lateral-tilt and longitudinal dynamics and both of them reside in two different places (target and host), this may induce some synchronization issue. However, the performance of the tracking of desired states of the NTTW vehicle is satisfactory.

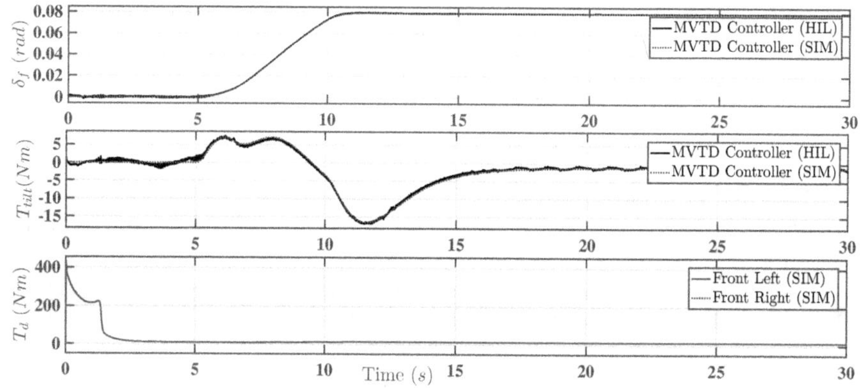

(a) Control inputs

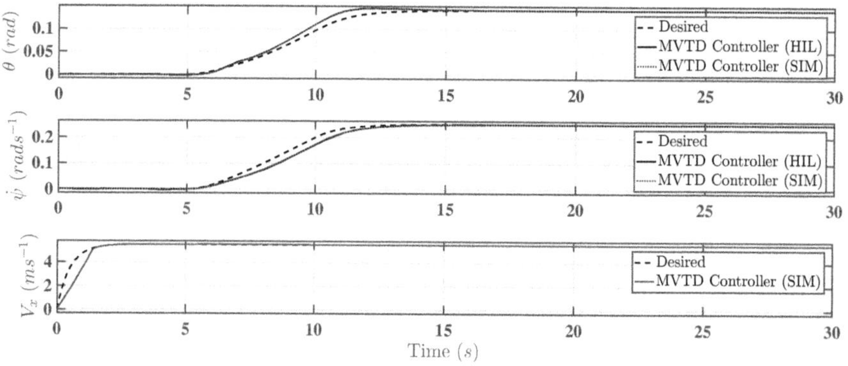

(b) States

Figure 10.5: HIL simulation results for MVTD controller during a step change

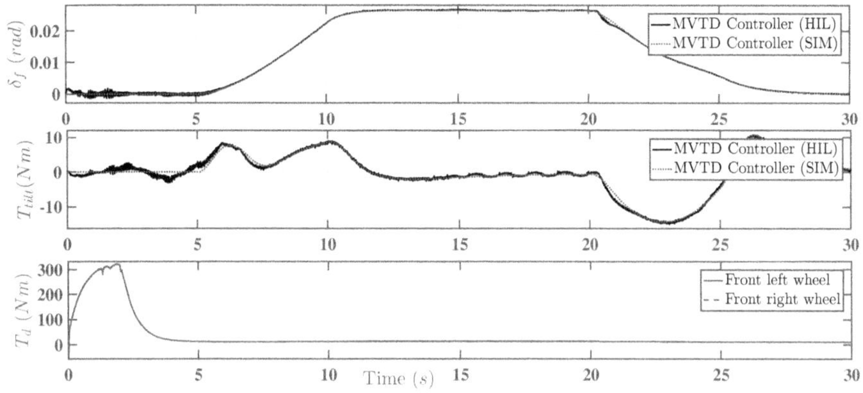

(a) Control inputs

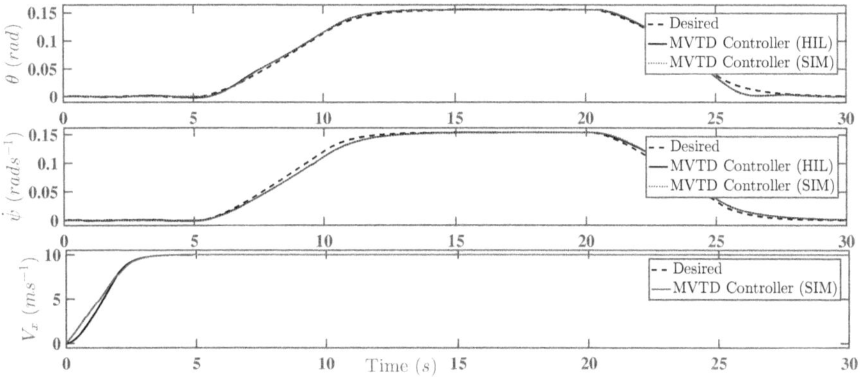

(b) States

Figure 10.6: HIL simulation results for MVTD controller during double lane change

Figure 10.7 and Figure 10.8 represent the results of the IM-SM controller as described in (8.15) when white Gaussian noise is added to sensors (as given in Table 7.3). Similar performance is observed in Figure 10.7 and Figure 10.8 where the HIL and the offline simulation results match with each other. Moreover, it is observed that the IM-SM controller is more robust to quantization noise as well as jitter occurring in HIL set up than the baseline and MVTD controller.

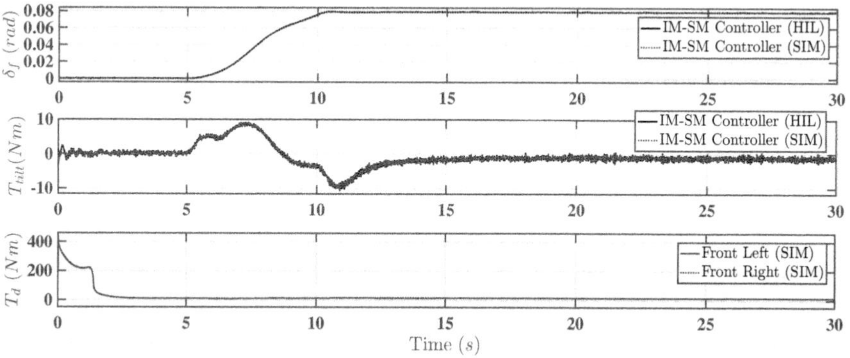

(a) Control inputs

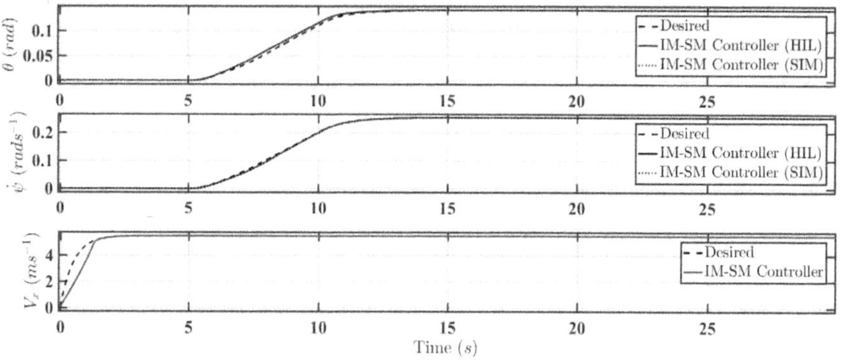

(b) States

Figure 10.7: HIL simulation results for IM-SM Controller during a step change

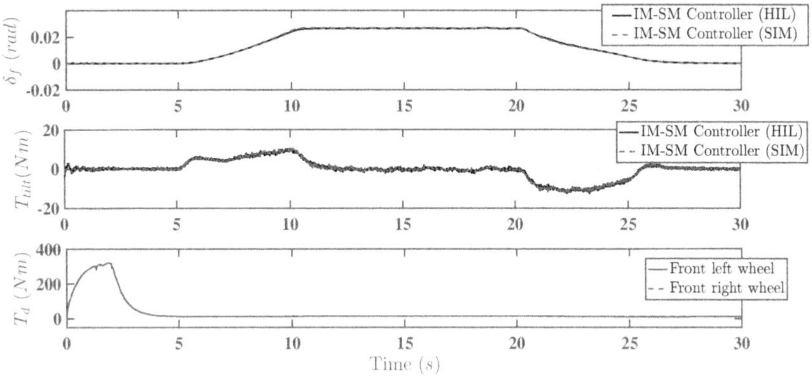

(a) Control inputs

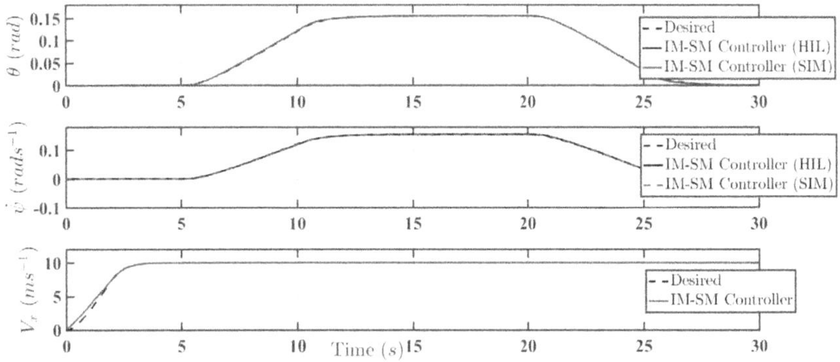

(b) States

Figure 10.8: HIL simulation results for IM-SM Controller during a double lane change

10.4 Summary

i. The HIL comprises of a host machine (simulating the NTTW vehicle model) and a target processor (housing the controller algorithm), which communicate with each other using suitable data acquisition box. This interaction between the host and the target enables the study of the control system in a more practical way. This HIL set up is used to validate that the proposed algorithms can be deployed on the embedded systems successfully.

ii. All three controllers have been programmed on STM board which is a low power embedded CPU based platform. The HIL setup deploys control calculations on to this CPU and the results match with the offline simulations. This indicates the feasibility of these complex multivariable controllers for practical deployments on automotive grade embedded ECUs.

iii. The HIL responses of all the controllers are compared with the simulation responses in each of the plots. It can be observed from the plots that the vehicle dynamics responses of simulation and HIL simulation match well with each other. Further, based on the results it can be inferred that IM-SM controller is more robust to quantization noise as well as jitter occurring in HIL set up than the baseline and MVTD controllers.

Chapter 11

Conclusion and Future Scope of Work

11.1 Introduction

The NTTW vehicle with in-wheel hub motors is highly suitable as an electric urban vehicle in congested cities. It combines positive aspects of a car (comfort, safety) and a motorcycle (low weight, dynamic driving performance). It outperforms conventional cars on the aspects of consumption of less space in traffic and in parking. However, its realization offers certain challenges in order to make it safe, comfortable and acceptable to the consumers. The vehicle dynamics in general consists of three translational motions (i.e. longitudinal, lateral and vertical) and three rotational motions (i.e. yaw, pitch and roll angles). The narrow tilting three wheeled vehicles has an additional degree of freedom for leaning of the vehicle body to achieve lateral stability.
The main achievements of this research are design and validation of an integrated chassis stability control system to attain the desired performance of the NTTW vehicle. A set of control objectives are drawn based on the functionality of the NTTW platform and typical vehicular stability conditions. The control objectives of the integrated control strategies are to track the commanded driver maneuvers in terms of yaw rate, tilt angle, longitudinal speed, and angular velocities of wheels, while maintaining vehicle dynamics stability. Mathematical modeling of this nonlinear, complex system is initially proposed in this work. Advanced control strategies are developed for desired performance of this unstable system. The proposed control strategies are subsequently validated through HIL simulations.

11.2 Concluding Remarks

The stability issues in a narrow tilting three wheeled vehicle are systematically studied and the boundary conditions of vehicle dynamics for stability are addressed. A robust control architecture

is formulated to achieve a balance of performance and power utilization.
Various test maneuvers, such as double lane change, constant radius turn, acceleration/deceleration while cornering, are simulated to gather significant information regarding different dynamic characteristics of the NTTW vehicle and the effectiveness of each stability control strategy. Further, each controller is tested for robustness against parametric variations and external disturbances, such as disturbance at longitudinal drag, road gradient, input and sensor noise. Based on the studies carried out during this thesis work the research achievements and the concluding remarks are elaborated below:

i. A detailed seven DoF non-linear NTTW vehicle model is formulated. The lateral-tilt dynamics consists of four DoF (V_y, $\dot{\theta}$, θ, $\dot{\psi}$) and the longitudinal dynamics comprises of three DoF (V_x, ω_{fl}, ω_{fr}). The lateral-tilt and longitudinal dynamics are interlinked through the non-linear tyre model (slip ratio, slip angle and tilt angle) and the vertical dynamics model (vertical tyre forces). It is observed that there is a strong coupling within the dynamics which adds to the complexities of the complete platform. It is also observed that the NTTW vehicle is open loop unstable.

ii. The objectives of tilt stability of the NTTW vehicle along with minimum steady-state tilting torque and zero steady-state value of perceived lateral acceleration (a_{per}) is met using baseline controller, which is based on classical PID control strategy. However, it is observed that the overall response of the baseline controller shows that the desired trajectory is not followed satisfactorily due to the steady-state error present in the other lateral state, i.e. the yaw rate. This implies that a control strategy is required which can simultaneously achieve the reference values of the yaw rate state as well as the tilt states of the vehicle while maintaining the desired longitudinal speed, which would essentially mean that the driver's command is being followed. Also, it is verified that the baseline controller is not robust to changes in the operating conditions of the vehicle due to its parametric variations and external disturbances.

iii. A multivariable control design based on optimal control strategy (LQR) is proposed for simultaneous control of the lateral-tilt states and the longitudinal states. Further, the LQR controller formulation allows the usage of a single state feedback gain across a range of vehicle speeds. Integral action is required to control the states simultaneously for MVTD. It is found that the input power requirement of MVTD is lower (for example, $\approx 50\%$ at 30 kmph for 32 m turning radius) than that of baseline controller. It is also observed that the MVTD controller follows the commanded trajectory closely whereas, the performance of the baseline controller for trajectory tracking is unsatisfactory. Further, MVTD controller is less sensitive to the effects of external disturbances, such as a sinusoidal disturbance added to the longitudinal drag. MVTD proves to be robust to speed changes and to the parametric changes (up to 20%) with respect to the mass, tilt moment of inertia and yaw moment of inertia of the NTTW vehicle. Thus, the MVTD control gives satisfactory performance in steady state and meets the required control objectives. However, it is observed that the transient state performance, such as variation in a_{per} and settling time of lateral-tilt states needs improvement.

iv. In practical scenario all the states are not available for feedback. Thus, the design of an estimator becomes an attractive approach to estimate the unmeasured state of NTTW vehicle. In this system, lateral velocity state in lateral-tilt dynamics is not measurable. On the other hand, the other states of lateral-tilt dynamics are measured using sensors. An EKF (extended Kalman filter) is designed for the lateral-tilt model to reconstruct all the four states of the system and it is tested under different conditions. It is observed that the EKF is able to provide noise-free estimates of all the lateral-tilt states of the NTTW vehicle satisfactorily. EKF is also found to be robust to parameter uncertainties.

v. A multivariable control design based on sliding mode control strategy (SMC) is proposed to simultaneously control the lateral-tilt states and longitudinal states. It is observed that the proposed controller (IM-SMC) and estimator (EKF) combination, is able to achieve the desired control objectives satisfactorily.

vi. The stability analysis of the proposed IM-SM controller is performed to ensure that it facilitates finite time convergence of surface to zero. The stability is evaluated by employing Lyapunov analysis and the limiting conditions for gain parameters for the controller are derived suitably based on Lyapunov criteria. With this approach it is verified that the closed loop system with the proposed controller is analytically stable.

vii. Comparison between two proposed multivariable integrated controllers is provided in quantitative as well as qualitative terms. It is observed that the peaks of the control inputs are lower and the response is faster in case of IM-SMC. Further, the variation in a_{per} is lower in IM-SMC than in MVTD controller which indicates better comfort for the driver and enhanced lateral stability of the NTTW vehicle. Thus, there is a significant improvement in the transient behavior of the NTTW vehicle due to IM-SMC. Consequently, the desired trajectory is followed in a better way in IM-SMC compared to MVTD controller. The input power requirements of the controllers are comparable in most scenarios, however, the MVTD requires more input power in the presence of external disturbances such as variation in vehicle resistance forces. Further, it is also observed that the NTTW vehicle state errors are comparatively higher in case of MVTD controller than the IM-SM controller. Also, the IM-SM controller rejects the external disturbances faster than the MVTD controller.

viii. An HIL setup has been created consisting of a real-time simulation of NTTW vehicle and controllers derived above. The Simulink Desktop Real Time executes the non-linear mathematical model of the NTTW vehicle, thus, representing the characteristics of the non-linear dynamics of the NTTW vehicle. The control algorithm (in the target block) and the vehicle model (in the host computer) interact with each other to form the HIL simulation. The controllers perform satisfactorily, hence validating the logic. All three controllers have been programmed on STM board which is a low power embedded CPU based platform. The HIL setup deploys control calculations on to this CPU and the results match with the offline simulations. This indicates the feasibility of these complex multivariable controllers for practical deployments on automotive grade embedded ECUs. It can be inferred from the results that the vehicle dynamics responses of simulation and HIL simulation match well with each other. Further, based on the results it can be inferred that IM-SM controller is more robust to quantization noise as well as jitter occurring in HIL set up than the baseline and MVTD controllers.

Hence, it is concluded that the proposed IM-SM controller is suitable to provide integrated dynamics stability for the NTTW vehicle. The robustness of the proposed controller is validated through rigorous simulation studies and further, by HIL simulations. Also, the multivariable control architecture allows the usage of combination of STC and DTC systems i.e. SDTC system, which basically means that the proposed strategy can be seamlessly incorporated with the current NTTW vehicle technologies.

11.3 Future Work

The following points suggest the directives for future work.

i. The mathematical model of the NTTW vehicle can be further enhanced by the inclusion of suspension and braking systems. The simulation analysis after the addition of these two systems would provide a more in-depth insight into the dynamics of NTTW vehicle. In present work, the actuators are modeled using first-order transfer function and their outputs are saturated to relevant magnitudes and hence the actuators are able to produce the required control efforts. However, in practical scenario, two stages are involved to provide the required control efforts for stabilization. Firstly the global control strategy determines the total control input that is required. Then, the control demands need to be mapped into relevant actuators to design their respective local control units while considering constraints, such as power etc. into account.

ii. An experimental prototype would enable to generate a substantial amount of data to further validate and refine the model and various parameters. A prototype for narrow tilting three wheeled vehicle could be developed as realistic vehicular parameters have been used during the analysis of this work. Ultimately, the performance and effectiveness of the proposed controllers are required to be analyzed and confirmed on the hardware prototype.

iii. The natural evolution of this work is the application of hierarchical control allocation methods to vehicles equipped with individual hub motors at each wheel. The functionality of the vehicle can be utilized further since the controller described in this thesis have implemented control of steering angles and traction/braking of the wheels. One area to improve further is to implement active torque vector control (TVC) that could be facilitated by in-wheel motors. The active torque distribution system provides active control of both traction and yaw moment. TVC may enable better performance of the NTTW vehicle during braking/deceleration while cornering, which has not been attempted for NTTW vehicle.

iv. The control of hybrid-electric vehicle drivelines along with dynamics stability analysis is an active research topic. This is another promising area where future research work could be explored.

v. NTTW vehicle requires multiple limits on states, control inputs, actuator operation. This calls for a control system that allows explicit treatment of various saturation limits. Furthermore, a complex control system is expected to work across larger envelope and have discrete and logical decision variables. The use of model predictive control strategy with mixed integer optimization formulation treatment can be explored as future work.